Safety and Justice Program

Evaluation of the Shreveport Predictive Policing Experiment

Priscillia Hunt, Jessica Saunders, John S. Hollywood

T0308357

The research described in this report was sponsored by the National Institute of Justice and was conducted in the Safety and Justice Program within RAND Justice, Infrastructure, and Environment.

This project was supported by Award No. 2009-IJ-CX-K114, awarded by the National Institute of Justice, Office of Justice Programs, U.S. Department of Justice. The opinions, findings, and conclusions or recommendations expressed in this publication are those of the authors and do not necessarily reflect those of the Department of Justice.

Library of Congress Cataloging-in-Publication Data is available for this publication.

ISBN: 978-0-8330-8691-4

RAND OFFICES
SANTA MONICA, CA • WASHINGTON, DC
PITTSBURGH, PA • NEW ORLEANS, LA • JACKSON, MS • BOSTON, MA
CAMBRIDGE, UK • BRUSSELS, BE
www.rand.org

Preface

Predictive policing is the application of statistical methods to identify likely targets for police intervention (the *predictions*) to prevent crimes or solve past crimes, followed by conducting interventions against those targets. The concept has been of high interest in recent years as evidenced by the growth of academic, policy, and editorial reports; however, there have been few formal evaluations of predictive policing efforts to date. In response, the National Institute of Justice (NIJ) funded the Shreveport Police Department (SPD) in Louisiana to conduct a predictive policing experiment in 2012. SPD staff developed and estimated a statistical model of the likelihood of property crimes occurring within block-sized areas. Then, using a blocked randomized approach to identify treatment and control district pairs, districts assigned to the treatment group were given maps that highlighted blocks predicted to be at higher risk of property crime. These districts were also provided with overtime resources to conduct special operations. Control districts conducted property crime–related special operations using overtime resources as well, just targeting areas that had recently seen property crimes (*hot spots*).

This study presents results of an evaluation of the processes in addition to the impacts and costs of the SPD predictive policing experiment. It should be of interest to those considering predictive policing and directed law enforcement systems and operations, and to analysts conducting experiments and evaluations of public safety strategies.

This evaluation is part of a larger project funded by the NIJ, composed of two phases. Phase I focuses on the development and estimation of predictive models, and Phase II involves implementation of a prevention model using the predictive model. For Phase II, RAND is evaluating predictive policing strategies conducted by the SPD and the Chicago Police Department (contract #2009-IJ-CX-K114). This report is one product from Phase II.

The RAND Safety and Justice Program

The research reported here was conducted in the RAND Safety and Justice Program, which addresses all aspects of public safety and the criminal justice system, including violence, policing, corrections, courts and criminal law, substance abuse, occupational safety, and public integrity. Program research is supported by government agencies, foundations, and the private sector.

This program is part of RAND Justice, Infrastructure, and Environment, a division of the RAND Corporation dedicated to improving policy and decisionmaking in a wide range of policy domains, including civil and criminal justice, infrastructure protection and homeland security, transportation and energy policy, and environmental and natural resource policy.

Questions or comments about this report should be sent to the project leader, John Hollywood (John_Hollywood@rand.org). For more information about the Safety and Justice Program, see http://www.rand.org/safety-justice or contact the director at sj@ rand.org.

Contents

Figures and Tables

Figures

Tables

Summary

Predictive policing is the use of statistical models to anticipate increased risks of crime, followed by interventions to prevent those crimes from being realized. There has been a surge of interest in predictive policing in recent years, but there is limited empirical evidence to date on whether predictive policing efforts have effects on crime when compared with other policing strategies. Against this background, the National Institute of Justice (NIJ) funded a predictive policing experiment conducted in 2012 by the Shreveport Police Department (SPD) in Louisiana; three districts used a predictive policing strategy to reduce property crimes and three control group districts continued with the status quo policing approach to reduce property crimes. Using qualitative and quantitative approaches, this report evaluates the processes, impacts, and costs of the SPD's predictive policing experiment.

The PILOT Program

Intervention Logic

The SPD's predictive policing effort, the Predictive Intelligence Led Operational Targeting (PILOT) program, is an analytically driven policing strategy using special operations resources focused on narrow locations more likely to incur property crimes. There are two overarching activities of PILOT: (1) using a predictive model to identify small areas at increased risk of property crime, and (2) implementing a prevention model to conduct policing interventions in the areas assessed to be at increased risk.

The underlying theory of the predictive policing program is that signs of community disorder, characterized by graffiti, public drinking, and rundown buildings, for example, and other data are associated with future crime risk. By focusing resources in areas predicted to experience increases in crimes, most notably by addressing disorder issues, the police can in principle deter and preempt crimes. (The SPD focused on property crimes in this case.) Figure S.1 illustrates the links between the activities and expected effects of PILOT, as well as the assumptions describing how the activities lead to the impacts in the intervention context.

Figure S.1
Logic Model for the Shreveport Predictive Policing Experiment

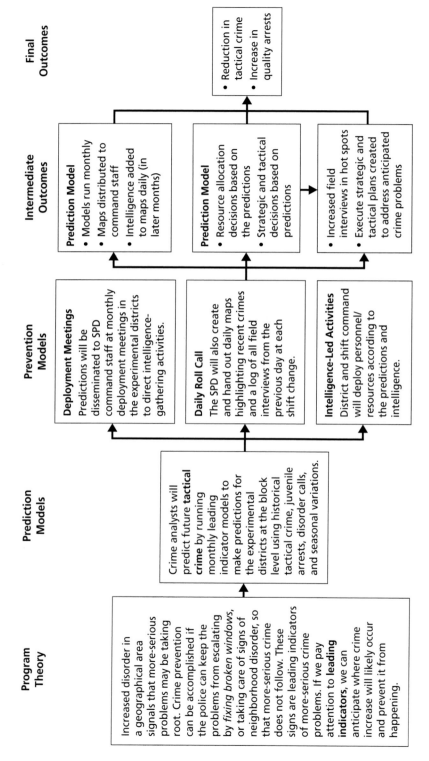

NOTE: *Tactical crime* is the SPD's term for six Part 1 property crimes: residential burglaries, business burglaries, residential thefts, business thefts, thefts from vehicles, and vehicle thefts.

Prediction Model

To generate predictions of future property crimes, the SPD performed multivariate logistic regression analysis for grid cells (400 feet on a side), covering each treatment district. Cells, or city blocks, with a predicted probability of 40–60 percent or 60+ percent of at least one property crime over the next month were marked in orange and red, respectively (see an example of a predictive map in Figure S.2). These orange and red squares can be thought of as *predicted hot spots*. Although not included in the original model, after several months, additional information was plotted on the maps to help inform daily decisions of which of the predicted hot spots to target with available resources. Plotted data included locations of recent crimes, 911 calls, field interviews, and recently targeted hot spots.

Prevention Model

As proposed, the prevention model used NIJ funding to support overtime labor assigned to the predicted hot spots. The specific strategies employed by the staff assigned to the hot spots in experimental districts were to be determined at monthly strategic planning meetings.

The control districts also used overtime funding to conduct special operations to target property crimes. Special operations were conducted both by district personnel (drawing on a pool of citywide funding) and the citywide Crime Response Unit (CRU) that acted in response to district requests. The principal difference for control districts was that special operations targeted small areas in which clusters of property crimes had already occurred—that is, hot spots were derived from conventional crime mapping techniques rather than from a predictive model. In addition, the only permitted special operations targeting property crime in the experimental districts were the PILOT interventions; experimental districts were not, for example, allowed to request CRU operations during the experiment.

Results on the PILOT Process

For the most part, all treatment districts adhered to the prediction model, as envisaged. Although there was variation over time after it was decided that the maps would be more useful if overlaid with situational awareness intelligence (e.g., locations of previous-day arrests and field interviews), this deviation from the initial program model was a progression based on what field operators requested. Similarly, the level of effort expended on the maps was consistent with the original model of weekly generation, although it was determined early on in the trial that it was only necessary to estimate the logistic regression models monthly, since predicted hot spots changed little over time, and later it was decided to update the maps daily with the situational awareness information.

Figure S.2
Example of a District's Predictive Map

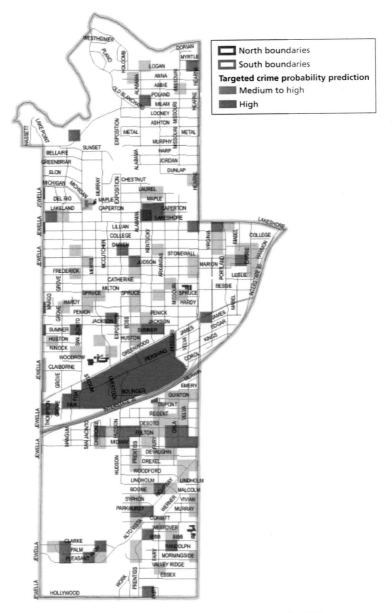

SOURCE: Courtesy of the SPD; reprinted from Perry et al., 2013.
RAND RR531-S.2

Conversely, treatment districts did not follow all aspects of the prevention model. Most important, the monthly planning meetings to set and maintain intervention strategies did not occur. These meetings were to be a key mechanism to ensure the prevention strategies were the same across police commands, and consequently to increase the statistical power needed for the impact analysis. Instead, the experimental districts made intervention-related decisions largely on their own. The resulting interventions all did assign overtime resources to predicted hot spots, as required. However, the strategies and levels of effort employed varied widely by district and over time. One command devoted many more resources to the intervention than the other command. Further, in general, the amount of resources declined significantly over time; we observed a reduction in the labor devoted per month from the start to the end of the evaluation period of more than 60 percent. The SPD made the decision late in the program to expand the experimental period months to take advantage of some of leftover funding; however, due to limited remaining resources and both planned and unanticipated leaves, staff reported not being able to operate the intervention much during the final months.

Table S.1 compares the two strategies and levels of effort employed. (Two experimental districts, A and B, were under the same area command and carried out the same strategy.)

Results on Crime Impacts

This study found no statistical evidence that crime was reduced more in the experimental districts than in the control districts.

Several factors that might explain the overall null effect have been identified, including low statistical power, program implementation failure, and program theory failure. The first is that the statistical tests used had low statistical power, given the small number of experimental and control districts, as well as low and widely varying crime counts per month and district in Shreveport.

The second factor, program implementation failure, is clearly seen in the treatment heterogeneity. We have noted low fidelity to the prevention model. In the first four months of the experiment, the level of effort was relatively high. During this period, we observed a statistically significant 35 percent average reduction in property crimes per month in the Command 1 districts (A and B) in comparison with their control districts. This study finds a statistically significant reduction in crime for all experimental districts compared with control districts during the same period, with virtually all of the reduction coming from Districts A and B. Yet in the last three months, when the level of effort was low, an increase in crime compared with the control districts is observed; however, the increase was not statistically significant. Conversely, we did not observe a significant change in crime in Command 2 (District C). Results are incon-

Table S.1
Intelligence-Led Activities, by Command

Element	Command 1 (Treatment District A and B)	Command 2 (Treatment District C)
Crimes targeted	Ordinance violation (and suspicious activity) perceived as preceding property crime by individuals with serious and/or multiple priors; individuals with warrants.	Narcotics, truancy, individuals with warrants (in general—specific strategies were more unit specific and ad hoc than in Command 1).
Staffing—directed patrol in predicted hot spots	Two patrol cars, each with two officers; one dedicated to driving and one to looking for suspicious activity and others to talk to. Cars do not respond to routine calls for service.	Two patrol cars, one officer in each. Cars respond to calls for service in addition to patrolling predicted hot spots.
Staffing—support	One dedicated sergeant to supervise; lieutenant and detective on call to respond to actionable information and developments. Average ratio of supervisor to officers: 1:5.	One supervisor available (not dedicated to PILOT only). Average ratio of supervisor to officers: 1:12.
Directed patrol—field interviews	Patrol units look for persons to stop who are violating ordinances or are otherwise acting suspiciously. During the stop, persons without serious records or warrants are told that police are looking for information about property crimes in the area and asked if they have information, along with how to provide tips. Persons with serious records or warrants were searched and arrested if applicable.	Patrol units look for persons to stop who are truant or are otherwise acting suspiciously. During the stop, persons without serious records or warrants are told that police are looking for information about narcotics crimes in the area and asked if they have information. Persons with warrants were searched and arrested if applicable.
Directed patrol—responses to property crimes in PILOT action areas	PILOT officers would canvas the area immediately around a crime that occurred in an orange or red cell, interviewing witnesses and neighbors as available about the crime to see if they had information. If actionable information was collected, they would call in the supervisor, detective, and other officers, if necessary.	When not responding to calls, PILOT officers patrol predicted hot spots. Officers may question people about narcotics trade.
Decisionmaking	The district lieutenant (and sometimes chief) and crime analysts decided on where to focus PILOT operations after reviewing the predictive maps augmented with the locations of recent crimes, field interviews, suspicious activity, and prior PILOT interventions. Decision criteria on which hot spots to target included the monthly forecasts, concentrations in recent activity, and a desire to not revisit places that had just been patrolled.	Lieutenant provided PILOT officers with maps and indicated target highlighted cells, with a focus on truancy and narcotics. Officers can decide strategy on a case-by-case basis.

Table S.1—Continued

Element	Command 1 (Treatment District A and B)	Command 2 (Treatment District C)
Key criteria	Reductions in Part 1 crime. Clearances of Part 1 crime. Quality stops and arrests (of persons with prior criminal histories, who were not in good standing with their community supervisor, etc.).	Reductions in drug offenses and property crime. Clearances of Part 1 crime.
Human resources policies	Selective recruitment of officers expressing interest and demonstrating specific skills; continued compliance with the elements above was necessary to continue being part of the intervention.	Recruit volunteers from any district.
Level of effort	4,062 officer hours (for 2 districts).	1,172 officer hours (for 1 district).

clusive as to exactly why the drop in crime occurred in the first four months of the experiment (since we have only one case of the Command 1 strategy), but it is clear that the two commands implemented different strategies to intervene in the predicted hot spots at different levels of effort and saw different impacts.

A third possibility is that the PILOT model suffers from theory failure—meaning that the program as currently designed is insufficient to generate crime reductions. We identify two possible reasons for the model theory failure. The predictive models may not have provided enough additional information over conventional crime analysis methods to make a real difference in where and how to conduct the intervention. Alternately, as noted, the preventive measures within predicted hot spots were never fully specified, much less standardized across the experimental group. This ambiguity regarding what to do in predicted hot spots may have resulted in interventions not being defined sufficiently to make an observable difference.

Results on Costs

Using administrative data on the time that officers devoted to property crime–related special operations and the time that crime analysts and personnel used to conduct predictive and prevention model activities, as well as the average pay by job type and standard estimates for vehicle costs (assuming that treatment and control districts had similar overheads), this study finds that treatment districts spent $13,000 to $20,000, or 6–10 percent, less than control districts. Whether these findings imply that the PILOT program costs less than status quo operations, however, requires assuming that, given the level of funding, the control group operated on average as the treatment group would have if not for the experiment. According to the SPD, this is a likely assumption because special operations are dependent on crime rates, and district commanders respond to increased crime by using special operations. Since the control group and experiment group were matched on crime trends, it is a reasonable assumption that on average they would have applied similar levels of special operations. This assumption would be invalid if there were differences in police productivity between the two groups (i.e., if the three districts of the control group used fewer man-hours on average to achieve the same level of public safety as the three districts of the treatment group). There does not appear to be a reason to presume that such differences exist. Furthermore, it seems unreasonable to assume that three districts would be permitted to consistently overuse police officer time compared with three other districts with similar crime rates. Since it was not possible to test this assumption, however, we generate best estimates, as well as lower- and upper-bound estimates.

Recommendations and Further Research

More police departments seek to employ evidence-based approaches for preventing and responding to crime. Predictive maps that identify areas at increased risk of crime are a useful tool to confirm areas that require patrol, identify new blocks not previously thought in need of intelligence-gathering efforts, and show problematic areas to new officers on directed patrol. For areas with a relatively high turnover of patrol officers, the maps may be particularly useful. However, it is unclear that a map identifying hot spots based on a predictive model, rather than a traditional map that identifies hot spots based on prior crime locations, is necessary. This study found that, for the Shreveport predictive policing experiment, there is no statistical evidence that special operations to target property crime informed by predictive maps resulted in greater crime reductions than special operations informed by conventional crime maps.

More-definitive assessments of whether predictive maps can lead to greater crime reductions than traditional crime maps would require further evaluations that have a higher likelihood of detecting a meaningful effect, in large part by taking measures to ensure that both experimental and control groups employ the same specified interventions and levels of effort, with the experimental variable being whether the hot spots come from predictive algorithms or crime mapping. In terms of specific strategies to intervene in the predicted hot spots, this study finds that one command reduced crime by 35 percent compared with control districts in the first four months of the experiment; this result is statistically significant. Given so few data points, conclusions cannot be drawn as to why crime fell by more in one command than in control districts. It is not possible to determine whether the strategy and the level of effort devoted to the strategy in Command 1 definitively caused the reduction. The strategy also raises sustainability concerns because neither the level of effort nor the crime impacts were maintained through the full experimental period; however, the strategy is worth further experimentation.

Acknowledgments

The authors would like to thank Susan Reno, Bernard Reilly, and all the officers and commanders at the Shreveport Police Department for the strong encouragement and assistance that they provided to support this evaluation. We are grateful to Rob Davis and Greg Ridgeway for contributions and advice in the early stages of this research. We would also like to thank our grant officer at the National Institute of Justice, Patrick Clark, for his support and advice during the course of this project.

Abbreviations

CRU	Crime Response Unit
CSA	cost savings analysis
FI	field interview
IT	information technology
NIJ	National Institute of Justice
PILOT	Predictive Intelligence Led Operational Targeting
RCT	randomized controlled trial
RTM	risk terrain modeling
SPD	Shreveport Police Department
VMTs	vehicle miles traveled

CHAPTER ONE

Introduction

Predictive policing is the use of statistical models to anticipate increased risks of crime, followed by interventions to prevent those risks from being realized. The process of anticipating where crimes will occur and placing police officers in those areas has been deemed by some as the new future of policing.[1] Since the early 2000s, the complexity of the statistical models to predict changes in crime rates has grown, but there is limited evidence on whether policing interventions informed by predictions have any effect on crime when compared with other policing strategies.

In 2012, the Shreveport Police Department (SPD) in Louisiana,[2] considered a medium-sized department, conducted a randomized controlled field experiment of a predictive policing strategy. This was an effort to evaluate the crime reduction effects of policing guided by statistical predictions—versus status quo policing strategies guided by where crimes had already occurred. This is the first published randomized controlled trial (RCT) of predictive policing. This report includes an evaluation of the processes, crime impacts, and costs directly attributable to the strategy.

Background on PILOT

Prior to 2012, the SPD had been employing a traditional hot spot policing strategy in which police special operations (surges of increased patrol and other law enforcement activities) were conducted in response to clusters—hot spots—of property crimes. The SPD wanted to predict and prevent the emergence of these property crime hot spots rather than employ control and suppression strategies after the hot spots emerged.

In 2010, the SPD developed a predictive policing program, titled Predictive Intelligence Led Operational Targeting (PILOT), with the aim of testing the model in the field. PILOT is an analytically driven policing strategy that uses special operations resources focused on narrow locations predicted to be hot spots for property crime.

---

[1]   For a full treatment of predictive policing methods, applications, and publicity to date, see Perry et al., 2013.

[2]   For information about the SPD, see http://shreveportla.gov/index.aspx?nid=422.

The underlying theory of PILOT is that signs of community disorder (and other indicators) are precursors of more-serious criminal activity, in this case property crime. By focusing resources in the areas predicted to experience increases in criminal activity associated with community disorder, police can in principle deter the property crimes from occurring.

As shown in Figure 1.1, the rationale behind PILOT was that by providing police officers with block-level predictions of areas at increased risk, the officers would conduct intelligence-led policing activities in those areas that would result in collecting better information on criminal activity and preempting crime.

To facilitate and coordinate deployment decisions and specific strategies within targeted hot spots, the proposed PILOT strategy depends on monthly strategy meetings. Day-to-day support in PILOT comes from providing officers with maps and logs of field interviews from the previous day at roll call meetings. Key outcomes were to include both reductions in crime (ideally, with the predicted and targeted hot spots not actually becoming hot spots) and quality arrests (arrests of persons for Part 1 crimes or with serious criminal histories).

Prediction Model

Following a yearlong planning phase grant from the National Institute of Justice (NIJ), the SPD developed its own statistical models for predicting crime based on leading indicators. The department experimented with a large number of models and found that there was a group of observable variables that was related to subsequent property crimes. The SPD reported predicting more than 50 percent of property crimes using its models (i.e., more than 50 percent of future crimes were captured in the predicted hot spots). The SPD generated predictions of future property crimes using multivariate logistic regression analysis for grid cells (400 feet on a side).[3] Cells, or city blocks, with a predicted probability of 40–60 percent or 60+ percent of at least one crime committed there over the next month were marked in orange and red, respectively (see an example of a predictive map in Figure S.2). Although not included in the original model, after some months additional information was also plotted on the maps, including recent crimes, reports of suspicious activity, field interviews, and reminders of which hot spots had been targeted previously.

Prevention Model

As proposed, the prevention model used NIJ funding to support overtime labor assigned to the predicted hot spots. The specific resource-allocation decisions and specific strategies to be employed by the officers assigned to the hot spots in experimental districts were to be determined at monthly strategic planning meetings. For day-to-day support, SPD crime analysts were to provide predictive maps showing the projected

---

[3] Squares of this size were chosen since they were approximately block sized.

Figure 1.1
Logic Model for the Shreveport Predictive Policing Experiment

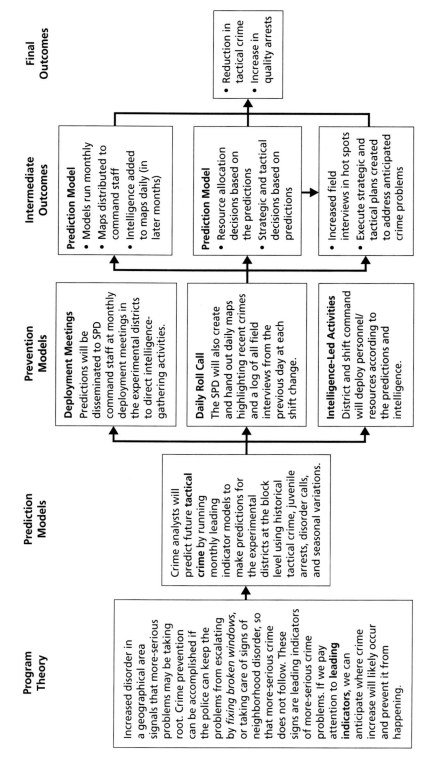

NOTE: *Tactical crime* is the SPD's term for six Part 1 property crimes: residential burglaries, business burglaries, residential thefts, business thefts, thefts from vehicles, and vehicle thefts.

hot spots during roll call meetings, along with spreadsheets describing recent crimes. The day's specific activities, consistent with the direction from the monthly planning meetings, were also to be decided during roll call.

The control districts also conducted special operations to target property crimes using overtime funding, which represents treatment as usual. Special operations were conducted both by district personnel (drawing on a pool of funding for Shreveport) and the citywide Crime Response Unit (CRU) acting in response to district requests. The principal difference for control districts was that special operations targeted small areas in which clusters of property crimes had already occurred—that is, hot spots derived from conventional crime-mapping techniques rather than from a predictive model. In addition, the only permitted special operations targeting property crime in the experimental districts were the PILOT interventions—experimental districts were not, for example, allowed to request CRU operations during the experiment.

The PILOT Experiment

Following the planning phase of the project, NIJ funded the department to conduct a field experiment of PILOT. The SPD implemented the experiment from June 4, 2012, through December 21, 2012, for a time span of 29 weeks. Although there are 13 districts in Shreveport, given budget, logistics, and crime-volume issues, the SPD decided it would be feasible to conduct the trial on six districts total—three control and three treatment districts.

A randomized block design was applied to identify matched control and treatment district pairs. Control districts conducted business as usual with respect to property crime–related special operations. The treatment-district chiefs, lieutenants, and officers were involved in a series of meetings with crime analysts to discuss when and how to implement PILOT.

Evaluation Approach

This report presents evaluations of the process, impacts, and costs of the SPD's PILOT trial. RAND researchers conducted multiple interviews and focus groups with the SPD throughout the course of the trial to document the implementation of the predictive and prevention models. In addition to a basic assessment of the process, we provide an in-depth analysis of the implementation, use, and resources over time of the predictive and preventive elements in the intervention model. It is hoped that this will provide a fuller picture for police departments to consider if and how a predictive policing strategy should be adopted.

Using administrative data from the SPD, this study also evaluates the impact of PILOT strategies on levels of property crime as a whole and by type of crime. Since the SPD implemented an experimental design, we test differences in crime between the control and treatment groups, and for robustness we also apply advanced program evaluation techniques (e.g., difference-in-difference methods). We also use findings from the process evaluation to test hypotheses for explaining program outcomes.

Lastly, this study evaluates the expenses of PILOT using a cost savings analysis (CSA). We use administrative data, and supplemental secondary data where needed, to calculate an estimated range of likely direct costs. Since crime trends were statistically similar between the control and treatment groups, and since procedures to acquire funding for special operations are dependent on the crime rates, the costs of property crime–related special operations in the treatment group are compared with the costs of property crime–related special operations in the control group. While the treatment group could not have spent more than the NIJ grant amount allocated to overtime operations, it could have spent less during the experimental period. This could occur because the grant period was longer than the trial period, and if all the funds for officer man-hours were not used during the trial period, one or more districts of the treatment group could continue with aspects of the program until the grant expired.[4]

In each chapter, we provide details of the methods and data used. The rest of this report is organized as follows: Chapter Two presents the process evaluation, Chapter Three details the impact of the PILOT program on crime, and Chapter Four presents the cost analysis. The final chapter concludes with a policy discussion and avenues for further research.

---

[4] Since the conditions of an experiment would no longer have been met (not all treatment districts receive the treatment), this study would not have included this later period in the evaluation.

CHAPTER TWO

PILOT Process Evaluation

This process evaluation describes the PILOT program implementation in Shreveport. Understanding how a proposed program was actually implemented is a critical part of evaluations, as it allows for understanding the source of effects on outcomes. Notably, if a program is not implemented as planned, it would be incorrect to ascribe any effects, whether positive or negative, to the planned program.

Note that the process evaluation takes no position on correct implementation, and rather describes what was planned, what actually happened in the field, and where there were opportunities and challenges. We present the results of the analysis for the prediction model and prevention models separately.

Methods

While there are a number of approaches to process evaluations, the one by Tom Baranowski and Gloria Stables (2000) is considered one of the most comprehensive process evaluation approaches (Linnan and Steckler, 2002). There are 11 key components, as summarized by Allan Linnan and Laura Steckler (2002, p. 8):

1. *Recruitment*—attracting agencies, implementers, or potential participants for corresponding parts of the program.
2. *Maintenance*—keeping participants involved in the programmatic and data collection.
3. *Context*—aspects of the environment of an intervention.
4. *Reach*—the extent to which the program contacts or is received by the targeted group.
5. *Barriers*—problems encountered in reaching participants.
6. *Exposure*—the extent to which participants view or read the materials that reaches them.
7. *Contamination*—the extent to which participants receive interventions from outside the program and the extent to which the control group receives the treatment.

8. *Resources*—the materials or characteristics of agencies, implementers, or participants necessary to attain project goal.
9. *Implementation*—the extent to which the program is implemented as designed.
10. *Initial use*—the extent to which a participant conducts activities specified in the materials.
11. *Continued use*—the extent to which a participant continues to do any of the activities.

We briefly assess the first seven components and, given their complexity and importance for transferability of results, conduct an in-depth analysis of the final four components (resources, implementation, and initial and continued use). Other than recruitment and continued use, all other components refer to activities during the trial period between June and December 2012.

Data

The process evaluation was conducted using mainly qualitative data collection. The RAND research team conducted in-depth interviews, field observations, and focus groups. Some administrative data are used as well. RAND researchers conducted comprehensive interviews with SPD crime analysts, captains, lieutenants, detectives, and patrol officers over a two-year period. Additionally, two sets of focus groups with police officers and crime analysts were conducted. One set of focus groups was conducted in November 2012, and a second, follow-up set was conducted in February 2013. Focus groups consisted of officers who had conducted operations in one district, separate from officers of other districts. This approach allowed for more-detailed, complete responses from various perspectives (e.g., captain, lieutenant, patrol officer, detective) per district. Focus groups with predictive analysts were separate from officers to reduce any bias in reporting on the process and experiences of the predictive policing experiment. The RAND research team also observed several roll calls and participated in five ride alongs.

All questions for the officers were open-ended and organized under the following themes:

- How they developed the prediction models
- How they developed the prevention models
- How the physical exchange of predictive maps, data, and information occurred
- What happened in the field
- What training was required
- What sort of resources (technology, labor, and equipment) were utilized

- What their perception of challenges, weaknesses, opportunities, and threats were
- How they plan to use what they learned.

Implementation of the Prediction Model

The first step of PILOT was to bring together a predictive analytics team to forecast the likelihood of property crimes in localized areas. The geospatial projections were based on tested models of factors that predict crime rates.[1] The predictive analytics team used statistical software to build and test regression models that estimate probabilities of crime, along with geospatial software to plot these estimated future probabilities of crime per geospatial unit onto maps. These initial models, as described by the SPD (Reno and Reilly, 2013), forecasted monthly changes in crime at the district and beat levels. The models were evaluated on their accuracy in predicting "exceptional increases"—cases in which "the monthly seasonal percentage change is > 0 and the district forecasted percentage change is > monthly seasonal percentage change [or] the monthly seasonal percentage change is < = 0 and the district forecasted percentage change is > 0." The models were estimated using several years of data.

Nonetheless, a change was made to the prediction model prior to the field trial. Following the October 2011 NIJ Crime Mapping Conference, the SPD team decided to change their unit of prediction to smaller geographic units. The change was largely due to concerns that predictions on district and beat levels were too large to be actionable, providing insufficient detail to guide decisionmaking. The team reported being further inspired by a briefing they saw on risk terrain modeling (Caplan and Kennedy, 2010a and 2010b), which predicts crime risk in small areas (typically block-sized cells on a grid). Thus, prior to the start of the experimental period, the SPD changed the methodology to predict crime risk on a much smaller scale—from district level to 400-by-400-foot grid cells.

The predictive analytics team tested three statistical approaches: risk terrain modeling (RTM) (Caplan and Kennedy, 2010a and 2010b), logistic regression analysis, and a combination of both. RTM identifies the cells at a higher risk based on the number of risk factors present in the cell.[2] Logistic regression analysis generates predicted probabilities of an event occurring based on indicator variables.

The SPD tested RTM using six density risk layers.[3] SPD staff also experimented with applying the count data used to generate the density risk layers directly into the

---

[1]   This was developed during Phase I of the NIJ grant.

[2]   Geospatial features correlated with crime.

[3]   The six density risk layers are probation and parole; previous six months of tactical crime, previous 14 days of tactical crime, previous six months of disorderly calls for police service, previous six months of vandalisms, and

logistic regression models (i.e., the risk layers were used as variables in a logistic regression analysis). Cells predicted to be at an elevated risk of one or more tactical crimes occurring next month were highlighted for action.

As shown in Figure 2.1, the logistic regression map tended to generate smaller, more-scattered hot spots than RTM, which displayed larger and more-contiguous hot spots. SPD staff also reported that RTM's hot spots tended to be already well-known to police in the area, whereas the logistic regression map produced some previously overlooked areas. As such, the SPD selected the logistic regression model, which identified cells at high risk of crime in the coming month using the following seven input variables:

- Presence of residents on probation or parole
- Previous six months of tactical crime (the six types of property crime included in the analysis: residential burglaries, business burglaries, residential thefts, business thefts, thefts from vehicles, and vehicle thefts)
- Tactical crime in the month being forecasted last year (which helps address seasonality)
- Previous six months of 911 calls reporting disorderly conduct
- Previous six months of vandalism incidents
- Previous six months of juvenile arrests
- Previous 14 days of tactical crime (which weights the most-recent crimes more heavily than crimes over the entire past six months).

While there have been concerns that predictive policing methods and interventions are discriminatory (see, for example, Stroud, 2014), the input factors used in the model strictly concern criminal and disorderly conduct information—no demographic or socioeconomic factors were used.

To validate the model, in early 2012, the SPD team compared model estimates of where crime was more likely to occur in July to December 2011 to where crimes actually occurred during those months. They found that cells colored as medium and high risk collectively captured 48 percent of all tactical crimes. The high-risk cells captured 25 percent of all tactical crimes, although if one extended the definition of *capture* to include 400-foot radii around the high-risk cells (near misses), then 49 percent of crime was captured.

In terms of predictive accuracy, the probability of a crime occurring in that month in cells labeled *medium* or *high* was 20 percent. For cells labeled *high*, the likelihood of a crime occurring in the month was 25 percent. The likelihood of a crime occurring within or around a 400-foot radius of a high-risk cell was 68 percent—in line with the predictions made by the earlier model, which was based on larger geographic areas.

---

at-risk buildings.

Figure 2.1
Example of RTM and Logistic Regression Displays for One District

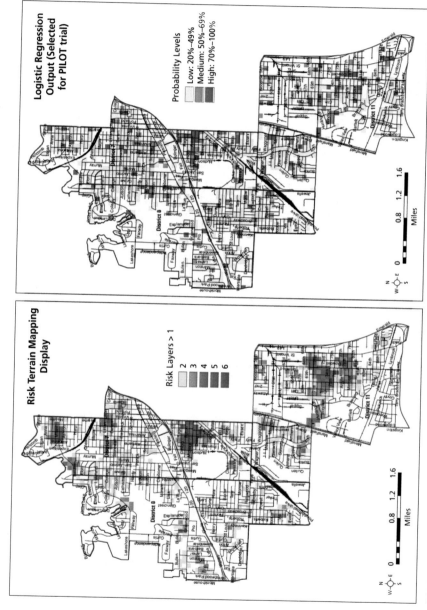

SOURCE: Courtesy of the SPD; reprinted from Perry et al., 2013.

NOTE: For RTM, the colors reflect the number of risk factors present in the cell. For the logistic regression output, *Medium* corresponds to a 40–60 percent predicted probability of a crime in the following month; *High* corresponds to a 60+ percent predicted probability of a crime in the following month.

RAND RR531-2.1

Implementation of the Prevention Model

Evidence indicates that the PILOT treatment varied across commands over time. There were three police commands involved in the trial by virtue of managing the control areas, treatment areas, or both:

- Command 1: District A and B (both treatment groups)
- Command 2: District C and D (one treatment and one control group)
- Command 3: District E and F (both control groups).

We describe what happened in each command separately since this was the source of variation.

Implementation in Command 1 (Districts A and B)

As described by officers, commanders, and the predictive analytics team, interventions in Districts A and B focused on building relationships with the local community in order to get actionable information for crime prevention and clearing crimes. There was a large emphasis on intelligence gathering through leveraging low-level offenders and offenses. Officers stopped individuals who were committing ordinance violations or otherwise acting suspiciously and would run their names through database systems.[4] If an individual had significant prior convictions, he or she would be arrested for the violation (as applicable). If the individual was on probation or parole, officers would check his or her standing with the parole or probation officers. For those not in good standing, a parole or probation officer was asked to come to the scene. Lastly, individuals with warrants were arrested. For those not meeting these criteria, officers stated that they gave these individuals a warning and were as polite as possible in order to note that they were trying to take action against property crimes in the area and to ask whether the individual had any knowledge that would be useful to police. Officers also noted talking to passersby in general about the SPD's efforts to reduce property crimes and asking if they had any potential tips. In the event that questioning led to potentially important information or an individual was arrested while in possession of potentially stolen goods, details were passed onto detectives.

According to operation officers in the districts, the PILOT strategy involved collecting more and better intelligence with the end goal of making quality arrests. Officers remarked that in past directed operations, the focus was on increasing arrests; however, with predictive policing, the focus changed to reducing the number of crimes. This appears to have been a significant change in mindset, because the metric for suc-

---

[4]   Most notably, Shreveport has an ordinance prohibiting walking in the middle of the street if there is a sidewalk. Interviewees noted this was not just for safety; it also reflects the perception that one of the few reasons for walking in the middle of the street if there is a sidewalk is to case houses and vehicles.

cess went from number of arrests to number of quality arrests, or arrests of individuals for Part 1 crimes or with multiple and/or serious prior convictions.

Officers reported that PILOT changed the amount of recent information provided per case. This occurred because PILOT officers at a crime scene asked more questions of victims, their neighbors, and individuals in the neighborhood, which is not normally done for property theft cases, as there is not enough time before officers have to respond to another call for service. Officers indicated that this was problematic for solving the case, because officers were essentially handing detectives a cold case when the detectives already had full caseloads.

Officers on PILOT operations said that they would conduct follow-up activities on good leads. In particular, when officers received potentially valuable intelligence during a field interview (with an individual possessing suspicious goods or providing a tip) or an incident report in progress, PILOT officers immediately notified the field supervisor and field detective. If necessary, other officers would be called onto the case. Officers indicated that they believed the response to be more coordinated than during normal operations.

Key elements of the command's strategy as reported to RAND are summarized in Table 2.1.

Implementation in Command 2 (District C)

Officers explained that the key difference of PILOT activities compared with normal special operations was that PILOT operations were a proactive strategy. Previously, directed patrols were conducted in response to crime spikes, such as recent increases in the number of residential burglaries. The earlier strategy was to apply x number of officers for y number of days in targeted areas. Operations were also conducted for known events (Fourth of July, Mardi Gras festivals, etc.), and officers from other districts were hired for those specific days and locations. PILOT operations, on the other hand, were focused on areas that may not have even experienced a previous crime spike.

The commander in this district did not have an overarching strategy to address PILOT predictions to the same extent as the commander for Command 1. Therefore, the prevention strategies were relatively ad hoc. That said, the PILOT officers did receive instructions to pursue suspicious activity, including, for example, questioning young people not in school. Officers indicated that, in general, they typically performed several key tasks during PILOT operations:

- Stopped and questioned juveniles committing truancy offenses
- Walked around apartment complexes and discussed criminal activities in area, particularly narcotics, with residents
- Visited people they know, especially parolees, probationers, and truants, to learn about criminal activities (largely drug activity) in the neighborhood.

Table 2.1
Intervention Strategy in Command 1

Element	Summary
Crimes targeted	Ordinance violation by individuals with serious and/or multiple priors; individuals with warrants.
Staffing—directed patrol	Two patrol cars, each with two officers; one dedicated to driving and one to looking for suspicious activity and others to talk to. Cars do not respond to routine calls for service.
Staffing—support	One dedicated sergeant to supervise; lieutenant and detective on call to respond to actionable information and developments.
Directed patrol—field interviews	Patrol units look for persons to stop who are violating ordinances or are otherwise acting suspiciously. During the stop, persons without records or warrants are told that police are looking for information about property crimes in the area and asked if they have information, along with how to provide tips. Persons with records or warrants for property crimes were searched and arrested if applicable.
Directed patrol—responses to property crimes in PILOT action areas	PILOT officers would canvas the area immediately around the crime, interviewing witnesses and neighbors as available about the crime to see if they had information. If actionable information was collected, officers would call in the supervisor, detective, and other officers if necessary.
Collection and analysis	Field interview cards were typed up and redistributed to district personnel daily; locations were also plotted on maps.
Situational awareness	Every day districts were provided with maps showing predicted hot spot areas for that month, along with locations of recent crimes, arrests, calls for service, field interviews, and prior police activity.
Decisionmaking	The district commanders and crime analysts decided on where to focus PILOT operations based on reviewing the maps described above. Decision criteria included the monthly forecasts, concentrations in recent activity, and a desire to not revisit places that had just been patrolled.
Key criteria	Reductions in Part 1 crime Clearances of Part 1 crime Quality stops and arrests—of persons with prior criminal histories, not in good standing with their community supervisors, etc.
Human resources policies	Staff participating in PILOT were reportedly enthusiastic about the interventions being conducted; continued compliance with the elements above was necessary to continue being part of the intervention.

District officers said that the main goal of PILOT activities was to protect citizens against burglary, with an emphasis on truancy and narcotics offenses that officers felt were causes or warnings for potential burglaries. Officers reported that there was good compliance with the intervention in the first months, but noted several logistical reasons for reduced man-hours. Some participating officers apparently started to feel that they were not performing well against performance measures, such as arrests, and that it was a waste of time and money (e.g., fuel) to drive to another hot spot and ask questions. The overtime incentives to participate in PILOT operations competed with other, more preferable overtime funds available to officers. Another practical problem arose: for the officers on overtime, there were a limited number of vehicles with air conditioning available during the summer. These issues led to difficulties in finding officers from within the district to conduct PILOT operations after the first month and resulted in an overall decline in man-hours devoted to PILOT. Similar to Districts A and B, District C also had to supplement with officers from other districts. Officers from Districts A and B also commented on the relatively lower-quality resources available to conduct PILOT operations in District C. Key elements of the command's strategy as reported to RAND are summarized in Table 2.2.

Table 2.2
Intervention Strategy in Command 2

Element	Summary
Crimes targeted	Narcotics, truancy, and individuals with warrants (in general, specific strategies were more unit specific and ad hoc than in Command 1).
Staffing—directed patrol	Two patrol cars, one officer in each. Cars respond to calls for service in addition to directed patrols.
Staffing—support	One supervisor available (not dedicated to PILOT only).
Directed patrol—field interviews	Patrol units look for persons to stop who are truant or are otherwise acting suspiciously. During the stop, persons without serious records or warrants are told that police are looking for information about narcotics crimes in the area and asked if they have information. Persons with warrants are searched and arrested if applicable.
Directed patrol—responses to property crimes in PILOT action areas	When not responding to calls, PILOT officers patrol red and orange cells. Officers may question people about narcotics trade.
Collection and analysis	Field interview cards were typed up and redistributed to district personnel daily; locations were also plotted on maps.
Situational awareness	Every day districts were provided with maps showing predicted hot spot areas for that month, along with locations of recent crimes, arrests, calls for service, field interviews, and prior police activity.
Decisionmaking	Lieutenant provided PILOT officers with maps and indicated target highlighted cells, with a focus on truancy and narcotics. Officers can decide strategy on case-by-case basis.
Key criteria	Reductions in drug offenses and property crime Clearances of Part 1 crime Command 1's change in officer evaluation criteria from total arrests to quality arrests appears not to have occurred here.
Human resources policies	Recruit volunteers from any district.

Activities in the Control Group (Districts D, E, and F)

In the control districts, as with all other districts not included in the trial, the crime analyst continued to provide typical crime statistics and bulletins of interest, and the information technology (IT) specialist continued to provide technical support. The usual crime statistics include both mapping of recent crimes and limited hot spot mapping using spatial kernel density methods.[5] Figure 2.2 shows an example of the type of crime map provided to control districts, which simply illustrates the locations of recent crimes of interest.

The reasons provided for limited kernel density mapping were that: (1) doing otherwise was considered time intensive and burdensome for the crime analysts, and (2) the resulting hot spots from kernel methods are fairly large areas with infrequent temporal variation, resulting in limited interest and use for patrol officers and leadership (since officers and leaders could quickly learn the locations of persistent hot spots). Similarly, the analytics team considered the special bulletins showing specific crimes in an identified hot spot rather ineffective for policing. Traditionally, the special bulletins would note crime spikes recognized over the course of several months at the head-

Figure 2.2

Sample of a Crime Map Provided to Control Districts

SOURCE: Reno and Reilly, 2013. With permission.
RAND RR531-2.2

---

[5] Kernel density methods use recent crime locations to generate maps that look similar to weather radar maps, only with more-brightly colored areas corresponding to places at higher risk of crime instead of precipitation intensity. Areas colored as hot spots typically are at least neighborhood sized. For more information on kernel density mapping (as well as other crime mapping techniques to generate hot spots), see Eck et al., 2005.

quarters level. The analytics team believed that the time lag limited the effectiveness of police to address the issue in the bulletin. Specifically, the time to identify a hot spot at the higher administrative level, send out a bulletin, and then send out an operations team means that the crime problem is likely to have moved.

To take action on emerged hot spots, districts have two options. The first is to request draws on a department-wide pool of overtime resources to conduct special operations. The second is to call on a citywide CRU that also provides short-term surge support. The analysis team, for example, noted that the control districts greatly increased their special operations in the last two months of the year compared with the experimental operations, in large part to combat anticipated spikes in stealing holiday presents.

The treatment districts did not have any access to the special operations pool or the CRU during the experimental period—all of their special operations were conducted strictly through PILOT support, as funded by the NIJ.

Summary

The intervention carried out in two of the treatment districts (both under one command, Command 1) may be considered more comprehensive and intensive. Officers were instructed to stop persons violating ordinances or acting suspiciously and question them about property crimes in the neighborhood; they were not to respond to calls. Individuals with warrants or violators of ordinances with serious or multiple priors were arrested. Command 2, in contrast, allowed officers to respond to calls and believed that a focus should be on drug offenses, so truants and suspicious persons were questioned about drug activity in the area. Individuals with warrants or suspected of committing drug offenses were arrested. The recruitment strategy differed as well; in Command 2, any officer volunteering for the overtime was assigned to PILOT, whereas Command 1 recruited officers with particular skills to a designated team. However, in the later months of the trial, both commands had to recruit outside of their districts. The two districts of Command 1 applied more than 4,000 hours, whereas the Command 2 district applied approximately 1,200 hours.

Results

Prediction Model
Program Fidelity

Broadly speaking, the prediction model initially implemented—maps of logistic regression analysis of 160,000-square-foot areas using seven indicators—was consistently applied throughout the duration of the trial. In months 1 and 2, the predictive analysis was run once each month, and maps were provided to officers with predictive grid squares. In these months, the SPD team advised districts to focus on highly clustered, high-risk cells.[6] During these first two months, there were discussions between officers and analysts to include more real-time information to help direct officers to where PILOT-related activities would be most effective. Furthermore, because the officers involved in PILOT activities changed from day to day, officers explained that information about the previous days' activities would help them to follow up with other officers' operations. Therefore, in months 3 through 7, the predictive analysis was still run once each month, yet daily updates of the following were marked on the maps and distributed to officers:

- Recent crimes
- Recent calls for service of interest (suspicious activity)
- Field interviews[7]
- Locations of PILOT operations.

As a result, the maps in months 3 through 7 of the field trial contained a much richer set of information—not only the monthly predictions of a tactical crime but also information from the day before. This type of almost real-time information is not routinely available to SPD officers. Due to a lack of data, it is unclear what impact this change in the program model had on officers' behavior and how officers actually used the situational awareness details to develop strategies for locations. We just have anecdotal reports from command and field personnel that the daily maps were useful in selecting focus areas for the day.

Examples of the two types of maps are shown in Figure 2.3: predicted hot spots from a day in months 1–2 are on the left, and on the right is a map from a day in months 3–7, which includes overlays of recent criminal activity, suspicious activity, and police activity. The later daily maps were referred to as *dashboards*, providing daily updated situational awareness information to the districts beyond the months-long forecasts.

As seen in the previous figure, maps included criminal activity, suspicious activity, and police activity. One indicator of the level of information provided in the maps is the number of field interview (FI) cards, which were filled out when an officer in the field made a stop of someone he or she felt was acting suspiciously. Table 2.3 shows the number of FI cards submitted by each treatment district during the intervention period. Prior to the experiment, FI cards were not systematically collected and therefore cannot be compared to numbers during the experiment. There was a greater number of FI cards submitted in the first three months, 140 to 274 FI cards per month, than later months, when the maximum number of cards submitted was 80 FI cards in

---

[6]   Returning to Figure S.2, the principal hot spot in the first few months in Command 1, for example, was the large group of orange and red cells directly below Interstate 20.

[7]   It should be noted that the SPD hired someone to enter all the field interviews into a computer to distribute them daily at roll call for this project. Normally, field interview cards are never electronically stored and are filed away in a cabinet.

Figure 2.3
Example District Prediction Maps

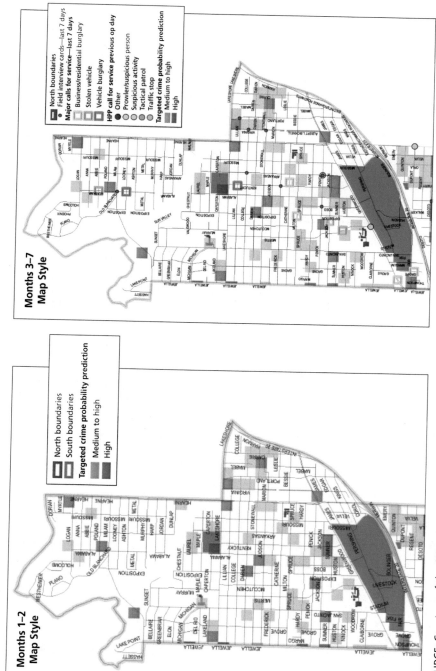

SOURCE: Courtesy of the SPD; reprinted from Perry et al., 2013.
NOTE: HPP = high-priority patrol.
RAND RR531-2.3

Table 2.3
Field Interview Cards

Intervention Month	By District			Total
	A	B	C	
June	93	1	46	140
July	163	8	103	274
August	103	9	46	158
September	54	13	2	69
October	23	8	1	32
November	54	15	11	80
December	18	7	8	33

a month. There appears to also have been a different amount of intelligence gathered across districts, with District A submitting up to 163 cards, District B submitting up to 15 cards, and District C submitting a maximum of 103 cards.

Program Dosage

In terms of the rate at which the predictive model was provided to officers (dosage), initially, the predictive maps for each district were generated multiple times per week. It was determined that running the model more frequently than monthly was unnecessary because there was already limited variation from month to month—hot spots in one month tended to remain hot spots the next month. Furthermore, the weekly estimated distributions were fairly large, indicating that weekly predictions were unreliable. The switch to running monthly logistic regression analysis was made early in the first month of the trial.

Prevention Model
Program Fidelity

The prevention model as proposed rested heavily on holding monthly planning and operations meetings at which district- and department-wide leaders would review the predictions and collectively develop strategies and resource allocations to address the predicted hot spots. However, these meetings did not occur, which left preventive activities largely at the discretion of the district commands. As a result, the two commands with treatment districts (Commands 1 and 2) conducted differing interventions that are difficult to reconcile as being part of the same preventive model.

Indeed, neither intervention can be said to be part of the proposed intervention model given that the proposed model centered on the monthly meetings to coordinate response strategies. The lack of multidistrict and department-wide coordination may also have led to some of the difficulties faced by District C in obtaining needed resources and developing consistent responses to predicted hot spots.

The two treatment commands reported conducting different types of interventions for PILOT, as well as managing the interventions differently. According to the predictive analytics team, they were directly involved with how the predictions would be acted upon in Districts A and B. They did not simply e-mail or send the maps to district chiefs, lieutenant, or officers. The crime analyst attended roll calls, and if a district's lieutenant was available and willing, the decisions of where to focus were jointly made. If the lieutenant was not available, the crime analyst made the subjective decisions of where to focus. In either case, decisions were based on a weighting of three factors: the predictive grids, intensity of recent activity, and where the PILOT units had recently been, in order to avoid focusing on the same places. In District C, the SPD crime analyst generated the same types of maps as shown above, but reported not typically having an active role in selecting PILOT activity areas.

The predictive analytics team and officers both noted that those conducting operations in Districts A and B typically were officers from those districts, whereas in District C, PILOT operations were typically conducted by officers recruited from outside the district. The predictive analytics team pointed out that this resulted in officers operating in District C having less familiarity with the area and the key players, as compared with District A and B PILOT officers.

The focus of operations differed. District C focused efforts on the narcotics trade, which the commander believed was the driver of the district's crime problems. Districts A and B concentrated officers on individuals with multiple and serious priors. Both districts encouraged arrests on warrants.

In terms of communication with district leadership, the predictive analytic team, officers, and the district lieutenant himself indicated that the lieutenant of District A and B was "field oriented" and directly contributed to PILOT operations by discussing strategies for particular locations on operation days, as well as directly supporting and supervising PILOT operations in the field. There was substantial communication between the crime analyst and Districts A and B's chief, lieutenants, patrol officers, and detectives. In contrast, in District C data flowed from the predictive analytics team to the lieutenant or officers, with much less direct involvement from the predictive analytics team; as noted, leadership in District C generally did not get involved with PILOT activities unless there was a pressing situation.

Program Dosage

The PILOT program was a redistribution of assets; it used additional man-hours in order to implement, record and report, and evaluate the trial. One way to understand how the implementation varied over time is to examine dosage—that is, the number of hours spent on the PILOT operations. There are two hypotheses about how officer hours would matter: (1) more time is needed up-front to put the program into place, and then that time can taper off (i.e., there are significant start-up costs), or (2) the time invested in the program needs to be steady to produce results.

In terms of the amount of overtime spent by officers in the treatment districts, approximately 5,200 hours over seven months for three districts, or 250 hours on average per month per district (approximately 60 hours per week), were devoted to PILOT operations. As Table 2.4 shows, the average is perhaps misleading, because there was a scaling down of PILOT operations from the third to fourth month; there was a 65 percent reduction in dosage from August to September.

In Districts A and B, the drop-off in effort was reportedly due to one key reason: lack of staff available to go out into the field. This was because of a combination of planned holidays, illnesses, and injuries; both Command 1 personnel and the crime analysts stated that having fewer available officers at both command and tactical levels degraded the command's ability to conduct effective PILOT operations. In District C, based on interviews, the drop-off appears to be due to insufficient resources (e.g., a lack of desirable vehicles); other, more-desirable overtime; and fewer officers available with any particular interest in testing the program.

Significantly, no one reported that reducing the overtime hours was reflective of the program working, which is what we would assume if there were program start-up costs. It is important to note that the initial planned experiment was only going to run from June to November 2012. The SPD continued the experiment in December, past

Table 2.4
Number of Man-Hours Devoted to PILOT, per District and in Total

Intervention Month	By District			Total
	A	B	C	
June	906.0	0.0	243.0	1,149.0
July	886.0	0.0	430.0	1,316.0
August	794.3	88.3	261.0	1,143.5
September	320.3	43.7	46.0	410.0
October	304.6	62.4	51.0	418.0
November	334.5	59.0	52.0	445.5
December	210.4	52.6	88.0	351.0
Total	3,756.1	305.9	1,171.0	5,233.0

NOTE: The man-hours for Districts A and B are SPD estimates, since these two districts ran joint operations under the same supervision, with most hours being allocated to District A for accounting purposes. Any differences are due to rounding.

the planned experimental period, since the department had some remaining funds from the NIJ grant.[8]

Summary

A summary of compliance to treatment, or program fidelity, is presented in Table 2.5. The predictive analytics team produced the maps and attended roll calls to provide and discuss maps and other intelligence gathered, as scheduled. The participants did not attend the planned monthly deployment meetings. According to qualitative data, participants did conduct intelligence-led activities in the places at greater risk of property crimes. However, because they did not meet monthly, the activities differed. Command 1 met with the predictive analytics team during roll calls; Command 2 received the maps but did not meet to discuss intelligence.

Command 2 applied a lower dosage of treatment than Command 1. Both commands reduced dosage in month 4 and onward. The quantity of predictive maps were provided for daily roll call as planned, although the predictions were run monthly (rather than weekly) and field intelligence was overlaid on maps daily.

Table 2.6 describes the components and a brief assessment of PILOT along 11 components of the process evaluation. Given the importance of knowing if and how a PILOT strategy could be rolled out more widely, we conducted in-depth analyses of implementation, use of information, and resources separately.

Table 2.5
Assessment of Compliance to Prediction and Prevention Models

	Treatment Compliance		
	Predictive Analytics Team	Command 1	Command 2
Prediction model activities			
Run monthly predictive models	Yes	N/A	N/A
Prevention model activities			
Monthly deployment meetings	N/A	No	No
Daily roll call maps and FIs	Yes	Yes	No
Intelligence-led activities (deployed according to maps)	N/A	Yes*	Yes*

* Activities differed.

---

[8] Specifically, SPD staff noted that since District C used less overtime than budgeted, there was some leftover funding used to extend Districts A and B's PILOT activities into December.

Table 2.6
Evaluation of PILOT Trial Process, by Component

Component	Assessment
Recruitment	The SPD received the police chief's buy-in and was successfully able to block randomize commands to receive the predictive analysis. The SPD did not, however, get permission to direct the district commanders on how to use the predictions. Therefore, there is high fidelity on randomization.
Maintenance	The SPD received funding from the NIJ to maintain data, which was all collected and made available upon request to the SPD. However, the department had a harder time keeping the district commanders engaged and implementing predictive policing prevention programming in the field over the entire seven months of the trial, as evidenced in the variation of man-hours over time.
Context	There is a large amount of seasonal variation in crime in Shreveport, so it is important for the generalizability of the trial for crime to be tested across each of these conditions. The trial was conducted during a period of relatively lower-than-mean, mean, and higher-than-mean crime rates in "medium" and "high" crime districts. Therefore, the context requirement was met.
Reach	The predictive analytics team was in contact with the district commanders on a regular basis, although the way they were integrated into the preventive programs differed across command. Regardless of the relationship between district commanders and the analytics team, the commanders reliably received the information at the proscribed intervals.
Barriers	Since this project was done internally, there were no barriers in reaching the district commanders. However, Command 1 was more engaged with the predictive analytics team than Command 2. The initial idea for the preventive intervention was for the predictive analytics team to have more involvement in the operations and an oversight board to ensure even implementation. Therefore, there were significant barriers in reaching the intervention targets as planned.
Exposure	From the interviews, it was clear that in Command 1, supervisors reviewed material with analysts and discussed operation strategies with officers. This translated into the field officers knowing about the intervention activities and being reminded of it during daily roll calls. In Command 2, the predictions were not reviewed with officers with the same frequency.
Contamination	Control groups were not exposed to the prevention model and did not receive maps or the same collected intelligence. However, they may have been exposed to the prediction models, since the SPD has COMPSTAT-style[a] meetings and districts had to recruit outside officers. It is conceivable that control districts communicated with treatment districts officers about their implementation strategies. However, not all officers in treatment districts were involved in PILOT, so if exposure did occur, it would have been limited.
Resources	Commanders reported having trouble getting officers for overtime, especially in the later months of the trial. The monetary resources are described in detail in Chapter Four.
Implementation	The predictive model was implemented with high fidelity; the preventive model was implemented with low fidelity. There is great variation in the implementation of the preventive model.
Initial use	As with many programs, there was a great deal of enthusiasm during the initial implantation, although there appears to have been more enthusiasm for the specific interventions in Command 1 than Command 2. Commanders and officers were more engaged, and more overtime hours were dedicated to the program.

Table 2.6—Continued

Component	Assessment
Continued use	Enthusiasm and/or energy may have waned over the course of the trial as it became less novel and more difficult to recruit officers for overtime operations, reportedly due to a combination of a lack of interest and a lack of sustainability, with commanders and key staff going on leave toward the end of the experimental period. Considering that this trial was only seven months long, it is troubling that the program was not able to gain traction and that interest waned so quickly.

[a] COMPSTAT (COMParative STATistics) refers to the presentation of crime statistics and analysis for the purposes of managing and coordinating police deployment decisions and strategies.

Key Implementation Challenges

Manual Generation of Predictions and Mapping Are Very Time Consuming

The field trial was a very time-consuming endeavor for the SPD crime analysis team; manually generating the crime predictions was reported to be an arduous task. There were two key reasons offered for the decision to not automate. First, the SPD is a relatively midsized department and did not have the budget to cover the costs of some of the crime prediction software programs. Second, the trial had not yet been conducted, and it was unclear what information would be most useful and relevant for officers. Overall, the Shreveport experiment was a large effort in terms of logistics—from preparing the maps to acting on the predictions to preparing reports—and the amount of effort was unsustainable in the long term (i.e., more than six months).

Recruiting PILOT Officers Is Difficult with Limited Resources

Officers in all districts reported that the key challenge of PILOT was to recruit officers. The PILOT strategy required gathering more information than usual; typically, officers have to be available to respond to calls, so they do not have the time to continue questioning victims, neighbors, and so on, whereas in all PILOT districts they were expected to do so. On the one hand, officers indicated that they could quickly identify when an individual was obviously not a viable suspect. On the other hand, when an individual was a potential suspect, officers stated that they needed more time for questioning because many individuals began the questioning by lying. Officers expressed that they spent a maximum of 30 minutes speaking with a suspicious person.

There was another reason why recruitment was a challenge to PILOT implementation: PILOT was an overtime strategy and thus required individuals to work additional hours. Districts A and B were able to implement a policy in which only those following the guidance on policing strategy were permitted to remain on the project and receive the additional pay. The commander indicated that the staffing strategy was to select a pool of officers with relevant skills and attitudes (e.g., communication, identification of patterns, and not driven by arrest-type performance metrics) that could conduct PILOT activities. However, Districts A and B faced multiple illnesses and injuries, so there were recruiting shortages and the districts borrowed officers from other districts. District C also had recruiting shortages, although for different reasons, which resulted in the need to recruit officers outside of the district; it was not in a position to implement a similar policy.

Coordinating a Singular Model for Taking Action Is Problematic

Although holding a monthly meeting was a key element of the prevention model, participants found it challenging. It is not clear whether the development of a collective group to decide on intelligence-led activities was not feasible, realistic, or ideal. There are a number of reasons why developing a group to build consensus on activities in two commands for half a year was difficult: commander management styles, ideologies on policing, neighborhood infrastructure design, and so on.

Observed Successes of PILOT Implementation

Improved Community Relations

According to officers, since the public saw so-called real criminals being inconvenienced and since police were conducting more follow-up questions after crimes, the public became more willing to provide additional information or call in with tips. This improved relationship with the public was evidenced, in part, by members of the public waving hello to patrol cars. Officers remarked that this did not happen prior to the operations. As a result, some officers in District A and B commented that the most positive outcome of PILOT was an improvement in relationships with residents. Prior to PILOT, officers felt that residents of areas with crime spikes had negative perceptions of the police and thus provided limited information on criminal activity. Officers in all districts believe that they obtained better and more crime tips due to increased intelligence gathering with residents following criminal activity, focused arrests of misdemeanor offenders with a serious criminal status, and improved discussions of crimes in the area with misdemeanor offenders in positive criminal standing.

Predictions Were Actionable (but Not Truly Predictive)

Officers in District C found the particular benefit of predictive maps to be that they provided a specific plan of where to conduct policing activities. Combined with data on the time of previous crimes in particular locations, officers knew when and where to work. Since the maps consisted of fairly small grid squares, officers were able to develop strategies for highly focused, specific areas, which allowed units to be more effective (e.g., quicker response times and more knowledgeable about potential suspects). Officers from all experimental districts discussed using the predictive maps to plan a strategy in terms of time of day and locations to focus on.

That said, one criticism of the predictive maps (notably from the District C commander) was that there were limits on how predictive the maps really were. The maps identified grid squares at increased risk of a property crime over the course of the entire next month.[9] The commander suggested that in future predictive policing efforts, the maps needed to be as specific as possible on when and where crime was likely to occur—otherwise the interventions would turn into directed patrols in hot spots, regardless of whether the hot spot came from crime mapping or a predictive algorithm.

An additional issue noted by officers was that the predicted hot spots changed little from month to month—for the most part, grid squares flagged one month would show up the next month. RAND team members, for example, observed officers on ride alongs (in Command 1) looking at newly distributed maps and remarking that they already knew where the hot spots were. They further stated that the core value of PILOT was in their command's focused patrols that were collecting better information on criminal activity and building better relations with the community, not in the predictive maps themselves.

Activities Improved Actionable Intelligence

When asked what they thought worked about PILOT, officers in all districts explained that the operation was an inconvenience to criminals, particularly those with priors for robbery, burglary, narcotics, or other property crimes. Furthermore, there was an overall belief that the skills and talents of PILOT officers, such as identifying patterns and working in teams, were instrumental in gathering relevant and timely information. Officers also reported that the maps made it easier to link a strategy with a location and with specific information; basically, they could concentrate their efforts. In the words of one officer: "The program increases our information, and anything limiting the bad guys' edge is helpful." Regarding the maps themselves, officers recommend overlaying reported and recorded crime in the previous day(s) (as was done later in the experiment). Since: (1) predictive hot spots were refreshed monthly and were stable over time; (2) recent activity may reduce or increase the likelihood of property crimes in the grid cells; and (3) the district only had resources to take action in a few hot spots per day, officers believed that the overlay of other recent data is useful for day-to-day decisionmaking.

---

[9] This is very different from popular perceptions of predictive policing, in which the software predicts almost exact times and locations of future crimes. See, for example, IBM's "Smarter Planet" advertising campaign, in which an officer shows up at a convenience store just in time to deter a robber.

CHAPTER THREE

PILOT Impact Evaluation

The PILOT strategy is to identify places at increased risk of experiencing crime and to both improve information gathering about criminal activity (leading to arrests for past property crimes and inconveniencing and deterring those considering future crimes) and maintain presence in those places.

In this chapter, we provide details of the methodology and results estimating the impact of PILOT on property crime.

Methods

Although this evaluation is based on activities during the implementation of the trial, the reliability of results depends on the robust selection of control and treatment groups preimplementation. Therefore, in addition to the estimation strategy and crime data, we present details on the approach to recruiting participants and tests of balance between control and treatment groups.

Identifying Control and Treatment Groups

As shown in Table 3.1, the SPD identified a block of four districts with higher violent and overall crime rates that consistently received special political attention and thus greater demand for police services. These four districts (District A, District C, District D, and District F) were assigned to the high-crime block, and the other matched districts were placed in another block.[1] The four districts in the high-crime block were randomly assigned to control and treatment groups (two in each group), and then two districts in the second block (middling crime) were randomly assigned, one to control and one to treatment.

The first randomization process selected Districts A and C for the treatment group and Districts D and F for the control group. The second randomization selected District B for treatment and District E for the control group. Within groups, experimental

---

[1] For this study, the districts have been relabeled with letters rather than their actual numbers for the sake of simplicity and to avoid overly focusing on district specifics.

Table 3.1
Block Randomization Data

District	Block	Tactical Crimes per Square Mile/ Three-Year Average	Disorderly Calls per Square Mile/ Three-Year Average	Drug Activity Calls/Three-Year Average	Group Assignment
A	1	170	1,150	263	Treatment
D	1	159	1,214	221	Control
C	1	168	947	242	Treatment
F	1	174	961	105	Control
B	2	84	639	153	Treatment
E	2	67	327	128	Control
G	2	44	204	135	Not assigned
H	2	37	202	98	Not assigned
I	2	51	215	62	Not assigned
J	2	29	134	54	Not assigned

District A was matched with control District D; experimental District B was matched with control district E; and experimental District C was matched with control District F. A map of the districts, Figure 3.1, shows where the districts of Shreveport are in relation to each other. The districts of the trial were all contiguous.

Testing the Match-Group Balance

The two groups were balanced in their pretreatment crime distributions, with no significant differences observed in their tactical crime, disorderly calls, or drug activity over a three-year period (see Table 3.2). In other words, the control group appears to be a good statistical match to the treatment group since crime trends over three years prior to the experiment were similar.

Empirical Model

The focus of the PILOT experiment was to address three tactical crimes thought to be influenced by police presence: residential, business, and auto-related thefts and burglaries. As such, the key outcome analyzed is the summation of these offenses into one overall crime type, *property crimes*. Additionally, the effect that PILOT has on each of the three crime types was analyzed separately. Given the seasonality of crime, we compare crime rates during the months of the trial with the same months in the previous year.

Figure 3.1
SPD Districts and Control and Treatment (Experimental) Groups

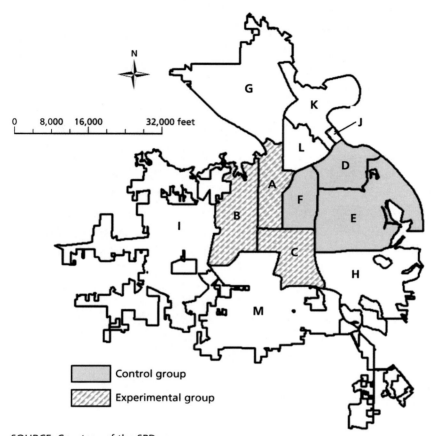

SOURCE: Courtesy of the SPD.
NOTE: Districts K, L, and M were never included because they are considered low-crime districts.
RAND RR531-3.1

Table 3.2
Testing Equivalence of Treatment and Control Groups

	Tactical Crimes per District (Three-Year Average)	Disorderly Calls per District (Three-Year Average)	Drug Activity Calls per District (Three-Year Average)
Control group	133.33	834.00	151.33
Treatment group	140.67	912.00	219.33
Difference	−7.33	−78.00	−68.00
t-statistic	−0.167	−0.258	−1.39

Given that the SPD implemented an RCT, we performed a series of statistical tests that compared control and treatment groups. For robustness, we also tested the difference between the crime change in the treatment groups with the crime change in the control group—that is, the difference-in-difference method (Bertrand, Duflo, and Mullainathan, 2004).

Impact Data

Using police-recorded data provided by the SPD, Figure 3.2 shows total property crime rates in the control and treatment districts in each month for five years before the intervention and during the intervention period in 2012. Generally, average property crime rates fell in Shreveport from 2007 to 2008 and stabilized thereafter. Shreveport crime rates appear to be seasonal, with more property crimes in June and July and relatively fewer in November and December.

Breaking down property crime into its relevant subtypes for the seven months of the trial and the seven months that preceded it, Figure 3.3 shows the changes in residential and business thefts and burglaries, as well as auto-related thefts (theft from autos, auto-accessory theft, and auto theft). The largest source of property crimes is residential thefts, followed by auto-related thefts. There were fewer residential and business property crimes during the seven months of the intervention than in the previous year for both the control and treatment groups. Auto-related thefts, however, increased in the treatment districts.

Figure 3.2
Property Crime Rates in Control and Treatment Districts

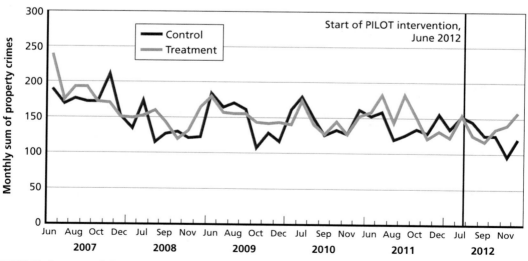

SOURCE: Courtesy of the SPD.
RAND *RR531-3.2*

Figure 3.3
Pre- and Posttrial Crimes in Control Versus Treatment Groups, by Crime Type

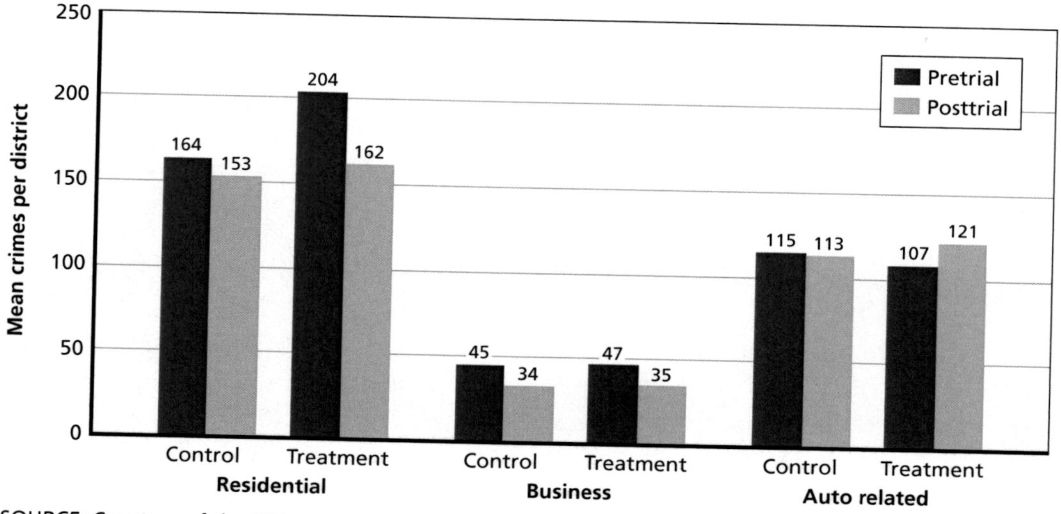

SOURCE: Courtesy of the SPD.
RAND *RR531-3.3*

Results

Main Results from the Randomized Controlled Trial

The following analysis examines the intervention period versus the same time period in the previous year. For example, it compares the crime of June 2011 with crime in June 2012, July 2011 to July 2012, and so on. There were approximately 24 fewer crimes per control district and 40 fewer crimes per treatment district. Results in the last column of Table 3.3 indicate that these reductions in crimes for each group are not statistically significant. The difference in crime reduction between the treatment and control groups (16 crimes) is also not statistically significant—meaning that PILOT did not reduce property crime. Since these statistics were derived from a RCT, it can be concluded that there is no statistical evidence that PILOT as implemented reduced crime or any of its subtypes.

Exploratory Analyses

There was no statistically significant impact of the program on crime overall, but it is unclear if that is because of a failure in the program model or a failure in the program implementation. Since the process evaluation clearly denoted treatment heterogeneity in the prevention model, it is also unclear which aspect of the program—the prediction model and/or the prevention model—could be responsible for program success or failure independently of the other. Readers should be cautioned to interpret the following exploratory analyses as just that—exploratory. These analyses cannot make any claims

Table 3.3
Testing Pre- and Posttrial Crime Differences, by Control and Treatment Group

	Pretrial	Posttrial	Difference
Treatment group (n = 6)	357.3	317.7	−39.7 (37.9)
Control group (n = 6)	324.0	300.3	−23.7 (37.9)
Difference			−16.0 (53.6)

NOTE: Robust standard errors in parentheses.

about causation and must be restrained in drawing conclusions about the PILOT program or predictive policing in general.

While the main aim of this outcome evaluation is to provide estimates of the overall effects of PILOT, it is also important to understand why a PILOT strategy may, or may not, reduce property crimes. Therefore, a series of follow-up analyses were conducted to give a more complete picture of changes in crime over the period investigated to help understand what is driving the overall average null result. These analyses are directed by our findings from the process evaluation. Since there was no variation in the prediction models across the three treatment districts, the following analyses focus on the differences in the prevention models utilized by the different commanders across the observation period. These differences are naturally occurring, so they can only tell us at how they correlate with changes in crime, not attribute causality.

When we examine the crime differences for the treatment and control groups monthly, the story is slightly more complicated, but the conclusion is the same: when averaged across the entire study period, there was no difference in the treatment and control conditions. The pattern over the duration of the intervention, presented in Figure 3.4, shows that crime fell by more in the treatment than the control groups in the first five months of the trial. This trend was reversed in November, when crime fell by 22 percent in the control group and increased by approximately 20 percent in the treatment group on the previous year.

Table 3.4 presents the differences between each treatment district and its matched control district in terms of changes in crime from the previous year. The results in the last two columns of the table confirm previous findings that PILOT had no overall effect on crime. The table provides results from an analysis of the short run (in the last column) and the total period of the experiment (in the second-to-last column). We define *the short run* as the first four months of the trial, from June to September, based on an analysis of the change in officer time devoted to PILOT and economic definitions of *the short run*.[2] By analyzing this short run, results reflect the time period

---

[2]   We considered two approaches to identifying the short run. First, we used a theory-driven approach, following economic theory that defines *the short run* as the period in which at least one factor of production (labor, capital, or land) is fixed and the others are variable. Reviewing the PILOT, this was the first three months of

Figure 3.4
Monthly Percentage Change in Crime on Previous Year, 2012

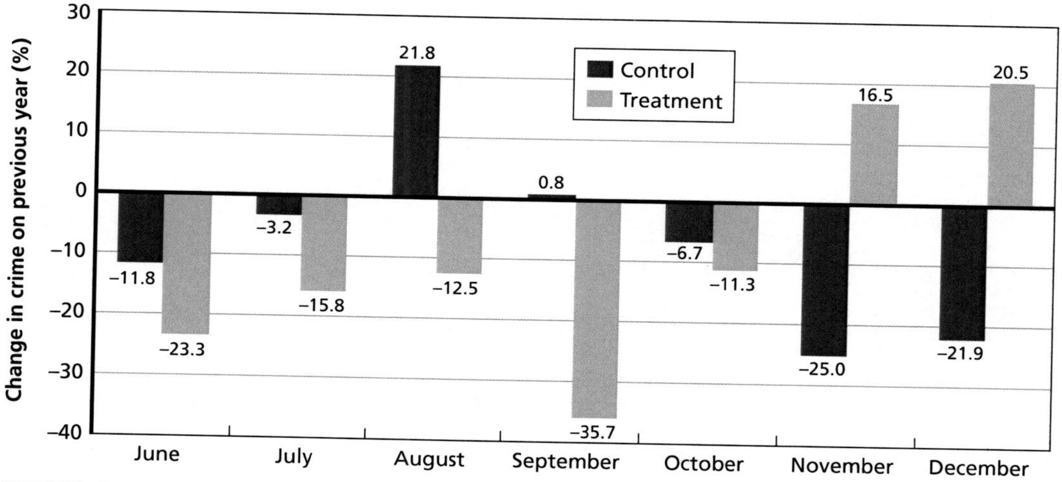

SOURCE: Courtesy of the SPD.
RAND RR531-3.4

in which labor resources were flexible and could more easily be increased or decreased as seen fit. This is also consistent with the interview evidence provided in the process evaluation in which districts, particularly A and B, were able to apply more officer time in the beginning of the trial but faced logistical problems after a few months.

There was a reduction in crime over the short run of the PILOT trial, as observed in Figure 3.4. The short-run reduction in crime was statistically significant, with an average of 13 fewer crimes per month per district. This finding varies by district; Districts A and B had a statistically significant fall in crime in the short run—19 and 28 fewer crimes per month, respectively—whereas District C did not. Districts A and B had similar changes in crime, but they each varied from District C in the short run; there was no variation between districts over the full period.

Readers should interpret these results with caution for three reasons. First, these results are not derived from the experimental design, so causation cannot be attributed to the program being evaluated. Second, multiple statistical tests were run on the same data, testing the same hypothesis, which means that several statistically significant findings are expected to be found by chance, even if no relationship truly exists. After

---

the trial. Second, we considered a data-driven model in which public safety (as measured by recorded crime) is a function of police force labor supply and tested associations between man-hours and crime. The data-driven approach identifies the short run as the first four months because of a lagged effect of police force labor on crime; this is not incompatible with the theory-driven approach, since one would expect a delay in the decrease of crime for increases in manpower from previous research (Caudill et al., 2013; Levitt, 2002; and Marvell and Moody, 1996). For more details on identification of the short-run, see the appendix.

Table 3.4
Exploratory Analysis of Variation in Property Crime, by Month and Districts

Treatment District and Matched Control District	Difference in Property Crime Levels (2011–2012)							Overall Average Difference (Across Entire Time Series)	Short-Run Average Difference (Across First Four Months)
	June	July	Aug.	Sep.	Oct.	Nov.	Dec.		
Treatment A v. Control A	−13	−28	−14	−22	−30	23	29	−7.9 (7.5)	−19.3* (8.2)
Treatment B v. Control B	−24	−25	−29	−32	8	17	32	−7.6 (8.1)	−27.5** (8.7)
Treatment C v. Control C	18	29	−1	−12	14	12	0	8.6 (5.9)	8.5 (8.3)
Average Difference	−6.3 (12.6)	−8.0 (18.5)	−14.7 (8.1)	−22.0 (5.8)**	−2.7 (13.8)	17.3 (3.2)	20.3 (10.2)	−2.3 (4.6)	−12.8* (5.9)
Comparing Commands[a]									
Command 1	−18.5	−26.5	−21.5	−27.0	−11.0	20.0	30.5	−7.7 (5.9)	−23.4** (6.8)
Command 1 v. Command 2	−36.5	−55.5	−20.5	−15.0	−25.0	8.0	30.5	−16.3 (9.1)	−31.9** (10.0)
Comparing Districts[a]									
Treatment A v. Treatment B	11	−3	15	10	−38	6	−3	−0.3 (10.3)	8.3 (7.0)
Treatment A v. Treatment C	−31	−57	−13	−10	−44	11	29	−16.4 (8.4)	−27.8** (8.8)
Treatment B v. Treatment C	−42	−54	−28	−20	−6	5	32	−16.1 (9.0)	−36.0** (8.9)

NOTE: Standard errors in parentheses.

[a] As compared with their matched controls.

** *p*-value < .01, * *p*-value < .05.

correcting for alpha inflation using the Bonferroni correction,[3] two findings remain statistically significant; the Treatment A/B versus Control A/B (*p*-value = .002) and the Treatment B versus Treatment C (*p*-value = .002) are significantly different in the short

---

3   This requires setting the *p*-value threshold for rejecting null hypotheses at 0.003 (.05 divided by 18 statistical tests).

run. More simply put: there is a difference in the effect in Command 1 versus Command 2 in the short run, and there is a difference in the effect of Treatment B versus Treatment C. Both of those models suggest that the intervention was associated with a larger crime reduction in Command 1 in the short run.

Third, by revisiting Figure 3.2, which illustrates the number of crimes by control and treatment groups between 2007 and 2012, it becomes apparent that there are several times during the past five years when the crime trends between the treatment and control districts deviate from one another as much as they do during the experimental period. It is, therefore, possible that the differences we observe in the first four months of the experiment are just part of the natural variation in crime.

An intriguing finding from Figure 3.4 is that in Command 1, year-over-year crime drops occurred in September and October, despite the fact that the labor hours devoted to the intervention fell from close to 800 to just over 300 hours per month. One possibility for the continued drops could be a lag effect. A second could be a positive effect from switching from simple monthly forecast maps to daily situational awareness maps that provide information on where recent activity of interest had occurred. A third possibility could simply be natural statistical variation.

Last, we considered the possibility of displacement and diffusion (spillovers) and criminal adaptation effects over time. Literature on spillover effects tends to find for diffusion—that is, additional crime reduction in surrounding areas.[4] Furthermore, there are geospatial characteristics of the experiment districts that would have limited the physical possibility of spillovers. In particular, one of the districts was bordered by a highway, and according to the SPD, the highway limits individuals' movement across its border. Also, note that most of the factors used in the Shreveport model—crime counts, disorder counts, presence of residents on community supervision—can change over the span of a few months in response to substantial variations in criminal behavior.

Limitations

A potential source of the overall null effect is that there was simply not enough statistical power to detect the true impact of PILOT. Results from the power tests, in Table 3.5, demonstrate that this experiment suffered from low power due to the combination of the low number of units involved (experimental $n = 3$, control $n = 3$) for a limited period of time ($T = 7$ months) with a low number of monthly property crimes in each district (the mean number of crimes equals 46 in the control condition and 51 in the experimental condition during the pretrial period) for a relatively wide variability of monthly crime within districts (the average standard deviation equals 9.5 in the control condition and 12 in the experimental condition during the pretrial period). While the experimental design is a huge advantage of this trial, the low power makes it very dif-

---

[4] See Bowers et al., 2011, for an analysis of 44 hot spot initiatives.

Table 3.5
Power Test Calculations

| | Treatment Group | | Control Group | | |
| | Mean (Standard Deviation) | | | | |
	Pretrial	Post-Pretrial Difference	Pretrial	Post-Pretrial Difference	Power
Test 1: A priori, given sample size	51.05 (11.82)		46.28 (9.48)		0.300
Test 2: Post hoc, difference between groups' difference in monthly property crime levels		−5.67 (16.05)		−3.38 (12.28)	0.079
Test 3: Impact needed to achieve power of 0.80		−16.15 (16.05)		−3.38 (12.28)	0.800

NOTE: Two-tail t-tests. Total sample size = 42, correlation = 0.260 (control group) and −0.041 (treatment group). Test 2 and 3: α error probability = .05. Test 1: β/α ratio = 6. α is the level of significance used in the test, and β/α is a measure of sensitivity to specificity.

ficult to detect anything other than a very large effect. Specifically, crime would have needed to fall by 30 percent (16.1 crimes per month on average, as opposed to 5.7) in the treatment groups to statistically identify the effect of PILOT. Recommendations for power range from .8 to .9, but the tests run in these analyses only had power ranges of .1 to .3.

Summary

This outcome evaluation examined changes in crime from a number of perspectives to provide a fuller picture of the effects of PILOT. Overall, the program did not result in a statistically significant reduction in property crime, as envisioned. This could be because the program does not work, the program was not implemented as intended, or that there was insufficient statistical power to detect the effect.

Several analyses were conducted to understand potential sources of this overall null effect by incorporating findings from the process evaluation. While the exploratory analyses suggest that there were different PILOT effects over time and across districts, the exploratory analyses cannot be causal due to the limited data available. Given the lack of multiple trials of the same intervention strategy, it was not possible to identify whether the decrease in crime observed in the early stages of the trial was due to PILOT prevention models or another factor, such as additional man-hours, normal crime variability, or another program or policing strategy. Evidence in the process evaluation indicated that the hours spent on the PILOT project dropped from an average of 1,200 overtime hours per month in months 1 through 3 to an average of just over 400 hours per month for months 4 through 7 of the trial, so there could be a prevention dosage effect. Furthermore, it also appears that one month—September—is a key driver of the positive findings in the short run, which could have simply occurred by chance and not be attributable to the program. Finally, it is also possible that PILOT's impact changes over time such that the intervention works better early on but the effect tapers off, resulting in a null effect overall.

CHAPTER FOUR

PILOT Cost Evaluation

Special operations funding in Shreveport is based on a criminogenic need—a district has to state the crime problem to acquire overtime funding. It is not the case that districts can conduct a property crime special operation because there is overtime budget available to do so; in the current system, a district can only access overtime funding for a special operation after a crime surge has occurred. One exception is for special events (e.g., Mardi Gras, winter holiday markets), in which case districts can conduct special operations for the expected increase in criminal activity rather than recent past crime surges.

One aspect of predictive policing that receives less attention, however, is the costs associated with implementing a prediction and prevention model. In an effort to bridge that gap, this chapter explores the type of direct costs that can be expected with a predictive policing program, and the chapter presents cost calculations and analysis.

Methods

In order to describe the costs of implementing a prediction and prevention model as done in Shreveport in 2012, this study conducts a CSA. Specifically, the direct costs of delivering property crime–related special operations in the treatment group are compared with the control group.

The cost estimations are centered on two main categories[1]—labor and equipment—because these are the two key resources used to implement both the prediction model (e.g., run models monthly, distribute maps to command staff, and collect and add intelligence data to maps) and the prevention model (e.g., make officer allocation decisions based on the predictions, make strategic and tactical decisions based on predictions, increase field interviews in hot spots, and execute strategic and tactical plans to address anticipated crime problems).

---

[1] There are two other main categories—land and capital—neither of which were features of PILOT.

Direct costs to the treatment and control districts were calculated using recorded data from the SPD and secondary data, where necessary.[2] *Direct costs* are defined as the expenses incurred as a direct result of implementing special operations for property crime. This study does not include indirect costs (e.g., overheads), because we assume that there is no differential between the treatment and control groups. We explored the possibility of including any potential additional revenues generated by the strategy (e.g., more paid citations); however, there was no strategy focused on revenue collection, which may have been needed in order to ensure substantial revenues from citations. Therefore, revenues generated by the strategy are likely to have been minimal. Furthermore, we determined that attributing such revenues to PILOT was not feasible or reliable.

Key Assumptions

The approach of this CSA is to compare direct costs of the treatment group and the control group using the actual recorded number of officer hours devoted to property crime special operations, estimates of analyst hours for the prediction and prevention model, estimates of vehicle expenses (which are based on the actual recorded officer hours), and IT costs. By comparing the control and treatment groups, the approach implicitly assumes that the control group is a good proxy for the labor and equipment expenses the treatment group would have incurred if not for implementing PILOT.

For this assumption to be valid, one key condition must hold: control districts operate the same as the treatment districts in the absence of the intervention. Therefore, using their resource utilization for the duration of the PILOT program is a good approximation for the resource utilization that would have been incurred in the treatment area if PILOT had not happened. If the block matching approach was done well, this condition should generally hold true; tests (presented in the process evaluation) show that the two groups are good matches.

According to interviews with the SPD, the approach seems reasonable because the crime trends were statistically similar between the control and treatment groups, and procedures to acquire funding for special operations are dependent on the crime rates.[3] This assumption would be invalid, however, if there were differences in police productivity between the two groups. For example, suppose the three districts of the control group could use fewer special operations man-hours on average to achieve the same level of public safety as the three districts of the treatment group. Then, even though they are statistically good matches for crime analysis, they are not good matches for resource analysis because for the same level of crime, the control group would always

---

[2] We do not include costs to administer the grant (e.g., reporting to the funder and meetings).

[3] A review of accounting and records systems indicated that the data were too limited to confirm or disprove this assumption. In particular, CRU and officer man-hours on property crime special operations for the two groups before the intervention were not available.

use fewer special operation police officer man-hours than the treatment districts. Yet we did not observe any reason to presume a fundamental difference in per-unit productivity between the control and treatment groups. While it seems unreasonable to assume that three districts would be permitted to consistently overuse police officer time compared with three other districts with similar crime rates, we do consider the possibility and, where relevant, take into account a potential range of productivity and provide lower- and upper-cost calculations.

In addition, it may initially seem that the availability of NIJ funding for the intervention affected the resources used, and therefore may not be informative of the costs that the experiment districts would have applied. On the one hand, the budget requested for officer time was based on previous levels of special operations and is indicative of the level of special operations that are conducted; although, as discussed previously, data were not available that separated CRU and officer time. Anecdotally, interviews with the SPD suggested that the nature of annual budgets means that there is limited scope for large changes annually in overtime man-hours. On the other hand, it is the case that there would have been an upper bound on the labor resources over the full period.

Table 4.1 presents an overview of the key assumptions and data used for the CSA, broken down by the two key elements—labor and equipment—of the policing strategies. Labor effort from officers, analysts, and administrators are considered in the cost calculations. According to interviews, the treatment group continued to receive some analysis that the control group received (e.g., bulletins); however, the exact amount of labor effort could not be determined. As such, the additional cost of performing tasks for the prediction and prevention model are calculated and attributed to the treatment group. An implication of this assumption is that if analysts and administrators did not continue providing treatment groups with the status quo forms of analysis and PILOT actually replaced some of these activities, then we will have overestimated labor costs for the treatment group; thus, we provide conservative cost savings estimates for PILOT.

The sensitivity of various assumptions, which provide the basis for the overall minimum, best, and maximum cost estimates, were tested. For more information on the assumptions and detailed calculations, see the appendix.

Cost Data
Labor Cost
Administrative data from the SPD show that the treatment group applied 5,233 officer man-hours to property crime special operations (see Table 4.2). A range of job classes, from probationary police officer to sergeant, was provided and resulted in an average hourly cost of $32 per hour. Combining the relevant hourly pay and number of hours, the total cost of officers for PILOT was approximately $168,000.

Table 4.1
Summary of Key Assumptions and Data for Cost Analysis

Factor	Data and Assumption, by Group	
	Control	Treatment
Labor expenses		
Police officer (man-hours and pay)	• Administrative records for district officers and CRU man-hours for property crime special operations • Administrative records for average overtime pay, by job class	• Administrative records for district officers man-hours for property crime special operations • Administrative records for average overtime pay, by job class
Crime analysts (man-hours and pay)	• No additional crime analysis performed	• Estimated range of man-hours to conduct predictive analytics and liaise with officers • Grant proposal average overtime pay, by job class
Equipment Expenses		
Vehicle (vehicle miles traveled and cost per mile)	• Literature estimates for police miles traveled per hour • Literature estimates for police vehicle cost per mile • Hours worked based on SPD data, with assumed range of hours traveled by job class	• Literature estimates for police miles traveled per hour • Literature estimates for police vehicle cost per mile • Hours worked based on SPD data, with assumed range of hours traveled by job class
IT (hardware and software, licenses and maintenance)	• Same as treatment group	• Same as control group

In addition, labor from analysts, systems administrators, and records personnel was needed to implement PILOT that went beyond what is typically done to support special operations. However, their time that was directly attributable to implementing PILOT, as opposed to reporting requirements of the grant, for example, was not recorded. Following assumptions based on interviews regarding the likely proportion of their time spent directly working to generate PILOT maps, implement maps in the field, and collect FI card information, this study estimates that roughly 300 to 800 analyst and administrative man-hours were directly attributable to conducting PILOT. (This range corresponds to one to three additional hours of administrative time per day.) Using SPD data on the average hourly pay of analysts and administrators, the total labor cost in the treatment group was approximately $176,000 to $193,000.

In the control group, administrative data indicate that 5,999 officer man-hours for property crime–related special operations were used. However, the job classes involved in the special operations were not available. Consistent with our assumption that the

Table 4.2
Labor Costs for Treatment Versus Control Groups

	Minimum	Best Estimate	Maximum
Control group			
Officers			
Total hours		5,999	
Weighted average hourly pay[a]	$32.59	$33.75	$34.11
Total	$195,498	$202,456	$204,616
Treatment			
Officers			
Total hours		5,233	
Weighted average hourly pay[b]		$32.04	
Additional analysts/records			
Total hours	291	582	873
Weighted average hourly pay[b]		$29.07	
Total	$176,110	$184,571	$193,032

[a] Using the average nonsergeant overtime hourly pay (probationary police officer to corporal IV) and sergeant overtime hourly pay and then weighting these two pay levels by the nonsergeant-to-sergeant ratio.

[b] Multiplying each job class hours spent by its average overtime hourly pay, and dividing by total hours spent.

two groups had similar productivity rates on average, this study applies the treatment group's average sergeant to nonsergeant ratio for the control group's best estimate. In order to address concerns with this assumption, we also apply the lower and upper ratios observed in the treatment group. Estimates denote a relevant average hourly pay of $32.59 to $34.11. This limited range is due to a combination of issues: the average nonsergeant (probationary police officer to corporal IV) hourly pay, $31.77, is only approximately 31 percent less than sergeant average hourly pay, $43.51, and the ratios observed still lean toward the predominant use of nonsergeants—that is, 13:1 (minimum), 5:1 (best), and 4:1 (maximum). Estimates suggest that the total labor expenses in the control group are $195,500 to $204,500. Regarding analyst and administrative labor, no additional labor was used for PILOT, since the control group did not receive the treatment and as was described earlier, analysts continued to provide the treatment group with some of the same types of analysis as the control group received (e.g., bulletins). Since it was not possible to know exactly how much time was spent on status quo analysis for each group, we err on the side of caution (i.e., underestimate cost savings of PILOT) and assume that both groups received the same status quo analysis, yet the treatment group also received predictive and prevention labor effort.

IT Equipment Costs

The type of strategy in PILOT requires one or more software packages that perform predictive analytics and plot the predictions on maps with highlighted spots. For the SPD, two software programs were used: analytical software (SPSS) to conduct logistic analysis for the likelihood of changes in crime rates and a geospatial software (ArcGIS) to plot the crimes rates in localized areas. The SPD said that preowned licenses for the software were used and no additional servers were used during the trial. Therefore, a cost difference of zero between the control and treatment groups for IT expenses was applied.

Vehicle Costs

Implementation of PILOT may have resulted in differential vehicle costs between the treatment and control groups. Administrative data on vehicle miles traveled for each of the groups was not available; therefore, survey data findings on average police vehicle miles traveled and police vehicle costs were used. The Federal Highway Administration finds that the average annual vehicle miles traveled (VMTs) by police in 2011 was 15,160 (American Public Transit Association, 2011). Vincentric (2010) reports that police vehicle expenses per mile are approximately $0.55, which includes depreciation, financing, fuel, maintenance, insurance, repairs, and opportunity cost. Using these values and assumptions about vehicle utilization, the vehicle expenses for property crime–related special operations were approximately $7,000 to $13,000 in the control group and $6,500 to $11,500 in the treatment group (see Table 4.3). The difference between the two groups is mainly due to the number of hours that patrol officers in particular worked on property crime–related special operations, and therefore takes into account potential group differences in resource utilization.

Table 4.3
Vehicle Costs by Control and Treatment Group

	Minimum	Best Estimate	Maximum
Control group			
Total VMTs	12,797	15,997	23,995
Total cost	$7,038	$8,798	$13,197
Treatment group			
Total VMTs	11,885	14,857	20,932
Total cost	$6,537	$8,171	$11,513

Results

Table 4.4 presents results of the CSA that compares the minimum, maximum, and best estimates of labor and equipment expenses for the treatment and control groups. Results show that PILOT cost approximately $13,000 to $20,000 less than the status quo. In other words, the treatment group spent 6 percent to 10 percent less than the control group to achieve a statistically similar change in crime over the seven-month trial period.[4]

In the minimum and best estimates, vehicle costs were 7 percent lower in the treatment than control group, and labor costs were 9 percent and 10 percent lower, respectively. In the maximum estimates, vehicle and labor costs were 12 percent and 5 percent lower than in the control group.

Discussion

This cost evaluation compared costs of implementing property crime special operations in order to provide another perspective on the effects of PILOT. The cost estimations include resources used to implement both the prediction and the prevention models. Overall, the program appears to have cost less to implement than the status quo opera-

Table 4.4
Summary of Cost Savings Analysis

	Minimum	Best Estimate	Maximum
Control group			
Labor	$195,478	$202,455	$204,600
Vehicle	$7,038	$8,798	$13,197
Total	$202,516	$211,253	$217,797
Treatment group			
Labor	$176,110	$184,571	$193,032
Vehicle	$6,537	$8,171	$11,513
Total	$182,647	$192,742	$204,545
Cost Savings	−$19,870 (−9.8%)	−$18,511 (−8.8%)	−$13,253 (−6.1%)

NOTE: Any differences are due to rounding.

---

[4] As these savings were used to continue PILOT in the districts after the trial evaluation period, these so-called benefits are not included.

tions as performed in the control group. Coupled with the outcome analysis that finds no statistical effect on crime, the implication of this analysis is that implementing PILOT as the SPD did over the trial period costs less without affecting crime.

A limitation of this result is that the approach uses the control group as a proxy for the resource utilization that the treatment group would have applied, given the level of funding available. This assumption seems reasonable since crime trends were statistically similar between the control and treatment groups and procedures to acquire funding for special operations are dependent on the crime rates. However, pretrial man-hours to verify similar levels of resource utilization prior to PILOT were not available for the control and treatment districts. Since it was not possible to test this assumption, we generated best estimates and allowed for a range of resource utilization to produce lower- and upper-bound estimates. For future cost analysis, it would be important to collect pre- and posttrial man-hours in both the control and treatment groups to make more-accurate, definitive statements about cost savings potential.

CHAPTER FIVE

Conclusions

This report presents evidence from a process, outcome, and cost evaluation of a seven-month field trial implemented in Shreveport, Louisiana, in 2012. The policing strategy—PILOT—included a prediction and a prevention model aimed to reduce residential, auto-related, and business property crimes.

Process, Outcome, and Cost Findings

Overall, the PILOT program did not generate a statistically significant reduction in property crime. There are several possibilities to explain the null results.

In part, the finding of no statistical impact may be due to a statistical issue—there were few participating districts over a limited duration, thus providing low statistical power to detect any true effect of PILOT.

The null effect on crime may also be due to treatment heterogeneity over time and across districts, which reduced the ability to statistically detect the impact of the program. There was high fidelity to the planned predictive model—the only difference was that the maps included other intelligence after some months—and dosage (level of effort) was relatively consistent over time (e.g., maps were generated for the daily roll call).

However, there was a lack of fidelity to the planned prevention model. There was more than a 50 percent decline in dosage after three months. As noted, there was a statistically significant reduction in crime in the control districts over the experiment's short run (June to September, with a p-value < .05).

Beyond dosage, participant districts did not meet monthly to discuss deployment decisions, which led to differing intervention strategies within hot spots. Districts varied in fidelity to daily roll call reviews of predictive maps and intelligence data.

The main challenges that may have affected fidelity and dosage sustainability to PILOT were:

- Lack of central coordination of a singular model
- Sustaining the near-daily manual generation of predictions and maps

- Limited resources to recruit officers for the overtime needed to carry out the interventions (the funding to pay for prevention labor in PILOT came from the NIJ)
- Avoiding burnout—sustaining operational focus and momentum, especially in the face of key absences (primarily observed in Command 1)
- Sustaining interest in the intervention (primarily observed in Command 2).

Another possibility is that the PILOT predictions are insufficient to generate additional crime reductions on their own. As noted, the SPD generated hot spot reports for both experimental and control districts over the course of the experiment, with the principal difference being whether the reports were reactive maps (showing recent crime) or proactive maps (showing the statistical predictions). The predictive model maps might not have provided enough new information over conventional crime analysis methods to make a real difference in where and how interventions were conducted.

A final possibility is that the PILOT preventative interventions are insufficient to generate additional crime reductions. As noted, the lack of monthly meetings meant that preventive measures within predicted hot spots were never specified and standardized across the experimental group. From an operational perspective, this left the experiment with not much of an option other than to devote more resources to predicted hot spots. This ambiguity on what do in predicted hot spots may have resulted in the interventions not being sufficiently different from typical hot spot surge responses to make an observable difference.

Although overall positive impacts were not identified statistically, officers across districts perceived the following benefits of PILOT:

- Community relations improved
- Predictions (the hot spot maps) were actionable; they provided additional information on the daily maps at roll call and summaries of FI cards, and predictions further assisted commanders and officers in making day-to-day targeting decisions.

Estimates of property crime special operations in the control and treatment groups indicate that the direct cost of PILOT operations in Shreveport in 2012 was 6 percent to 10 percent less than status quo operations.[1] This range is based on considerations that the control group may have had differential resource utilization levels than the treatment group would have used, if it had not been involved in PILOT.

---

[1] *Status quo operations* refers to hot spot mapping of past crimes and deploying patrols reactively to those crime hot spots.

Implications for Policy and Future Research

More police departments seek to employ evidence-based approaches to preventing and responding to crime, and predictive maps are a useful tool to confirm areas that require patrol, to identify new blocks not previously thought in need of intelligence gathering efforts, and to show problematic areas to new officers on directed patrol. For areas with relatively high turnover of patrol officers, maps showing hot spots may be particularly useful to help educate new officers about their areas.

What is not clear, however, is whether distributing hot spot maps based on predictive analytics will lead to significant crime reductions compared with distributing hot spots maps based on traditional crime mapping techniques. In the case of the Shreveport predictive policing experiment, we did not see statistical evidence to this effect.

More-definitive assessments of whether predictive maps can inherently lead to greater crime reductions than traditional crime maps would require further evaluations that have more power, gained in large part by taking measures to ensure that both experimental and control groups employ the same interventions and levels of effort, with the experimental variable being whether the hot spots come from predictive algorithms or crime mapping. Below are some potential elements to include in future predictive policing experiments:

- Test the differences between predictive maps and hot spots maps. What are the mathematical differences in what is covered? What are the practical differences in where and how officers might patrol?
- Carefully consider how prediction models and maps should change once treatment is applied, as criminal response may affect the accuracy of the prediction models.
- Understand what types of information practitioners need to tailor their interventions to the hot spot appropriately.
- Understand the difference between police officer and administrator predictions (based on experience) and predictive maps (based on statistics), and how the two might be combined.
- Understand how preventive strategies are to be linked with predictions.
- Check on how preventive strategies are being implemented in the field to ensure fidelity.

On a more fundamental level, it is worth considering if it is possible for some sort of predictive product to enable interventions that are fundamentally advanced beyond the sorts of directed patrols employed in PILOT. The commander of District C noted that unless predictive products could provide very high levels of accuracy, literally specifying where and when specific crimes were highly likely to occur, the predictions were just providing traditional hot spots, and interventions could only be typical hot spots interventions (e.g., focused patrols).

The promise of having highly accurate predictions has captured the public's imagination. For example, IBM ran a "Smarter Planet" campaign, which showed an officer driving to a convenience store just in time to meet a would-be robber. The reality is different—predictive maps to date have shown what we would consider incremental improvements in accuracy over traditional crime mapping methods at capturing future crime in predicted hot spots. This reflects the fact that traditional hot spot mapping is inherently predictive, assuming that future crimes will be concentrated where past crimes have been, which, based on recent RAND research, is usually reasonable. At the same time, the specific times and locations of crimes continue to be highly uncertain, with a great deal of seemingly random noise in exactly when and where a given criminal will choose to commit a crime, placing real-world limits on the accuracy of predictive models. See Perry et al. (2013) for a detailed discussion of predictive policing hype versus reality.

Thus, at least at the current time, predictive models will be generating what are essentially hot spot maps, albeit with more accuracy. A key research question, then, is whether it is possible to generate analytic products, including geospatial predictions, but adding additional information that would be more useful than traditional hot spots. The results from Shreveport do provide one intriguing possibility. As noted, declines in crime rates in Command 1 continued for two months after dedicated hours fell from more than 750 to around 300. This may be due to the delayed effect of police force levels since our analysis; other literature has demonstrated the delayed impact of police (Marvell and Moody, 1996; and Levitt, 2002). However, the timing also matches when the Shreveport crime analysts began generating daily maps that provided information about activities of interest (crimes, suspicious activity, field interviews, and recent police activity) in addition to just predictive maps.

Regarding linking prevention strategies with predictive maps, it is still ambiguous what works. During the first four months of the experiment, Command 1 (Districts A and B) experienced a 35 percent reduction in property crime compared with the control, which is when overtime hours dedicated to the project were comparatively high. The observed effect exceeds the norm; Braga, Papachristos, and Hureau (2012) find a smaller average treatment effect for hot spot interventions. It should be noted that Command 1's strategy allows for acting on hot spots generated through traditional crime analysis as inputs; it does not require a statistical model. While the command was not able to sustain the strategy throughout the experimental period, and it is not possible to say definitively that Command 1's intervention strategy and the level of effort devoted to it caused the short-term effect, it may be worth further experimentation.

Technical Details

Delayed Influence of Police on Crime

A statistically significant fall in man-hours occurred in the treatment group. This change in dosage may have been an important source of variation throughout the trial, which is examined in the body of the report. This event required the assessment of the association between police man-hours and levels of crime in order to determine the relevant short-run periods of analysis.

Research shows that there is a lagged effect of police on crime, particularly property crime (Caudill et al., 2013), in which the levels of police presence in previous periods have an effect on future levels of crime. Given this, the data were tested for the presence of a lagged relationship between police man-hours and crime. The readers should be cautioned that these tests are performed on a small number of observations ($n = 14$), and there is no model for causal inference; this is exploratory in order to understand if our data follow other evidence on police effects on crime.

Figure A.1 provides a visual representation of the relationship between past-month levels of manpower and current-month levels of crime. There are few observations and thus low power to detect an association, yet there may be a pattern in which up to a point (e.g., nonlinear effect) increasing man-hours is associated with reductions in crime in the following month.

Testing for simple correlation and considering a two-period lag as well (see Table A.1), there is a weak association between police man-hours in previous-month and current-month crime. While this is consistent with other research, a causal explanation cannot be inferred. It is simply indicative that there may be a delay that needs to be taken into consideration for the short-run analysis.

This is an important avenue for further research because it appears to be a central issue in the prevention model. Generally, treatment districts reduced resources over time, as shown in Figure A.2. In particular, there was a large decline in month 4 (September) of the trial. Assuming that CRU unit man-hours were consistent each

Figure A.1

Scatterplot of Group-Level Property Crimes and Previous-Month Man-Hours

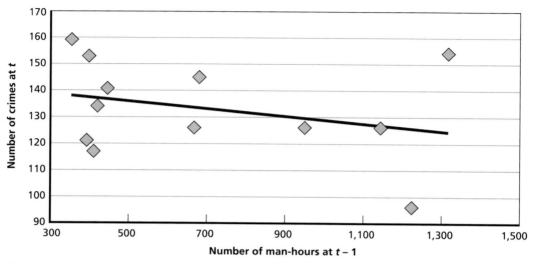

NOTE: n = 12, for 2012 data only.

RAND RR531-A.1

month,[1] the control and treatment groups appear to have had different labor utilization patterns. Whereas the treatment group maintained nearly the same number of hours in the months before September and then a lower, yet similar, number of hours from September onward, man-hours in the control group varied quite substantially from

Table A.1

Pairwise Correlation of Group-Level Officer Man-Hours and Property Crime

	crime,t	crime,$t–1$	crime,$t–2$	hours,t	hours,$t–1$	hours,$t–2$
crime,t	1					
crime,$t–1$	.2662	1				
crime,$t–2$	–.1050	.1062	1			
hours,t	–.0274	–.3165	–.0873	1		
hours,$t–1$	–.4736*	.2892	.2683	.0454	1	
hours, $t–2$	–.4843	–.5214	.1689	.3720	.2506	1

NOTE: n = 14.

* $p < 0.12$.

---

[1] Data on CRU man-hours were only available for the entire June–December period, so for monthly man-hours, we divided the total hours by the number of months.

Figure A.2
Monthly Man-Hours for Treatment and Control Groups, 2012

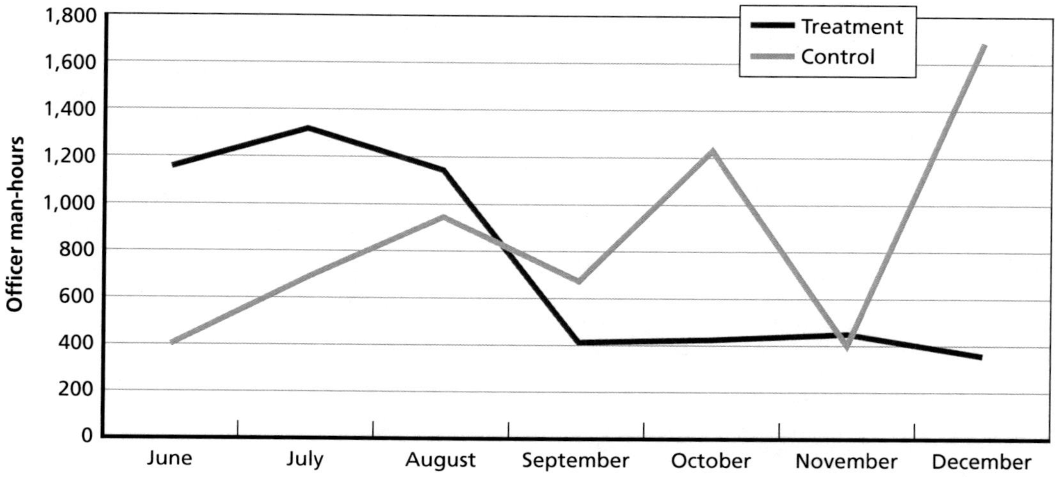

month to month. A two-group *t*-test finds that the mean man-hours before September and after in the treatment districts were statistically different (*p*-value = .002); the difference in man-hours was not statistically significant for the control districts (*p*-value = .39).[2]

Details of the Cost Analysis

Labor Costs

The SPD maintained records on the number of man-hours devoted to PILOT operations in each of the treatment districts and property crime–related special operations in the control group as a whole. Table A.2 presents the number of hours in total and by job class, where available. As seen in the table, the control and treatment group used approximately 6,000 hours and 5,230 hours, respectively, for property crime–related special operations. On average, that equates to approximately 13 hours per day per district and 11 hours per day per district for the control and treatment groups, respectively.[3]

Whereas all the treatment districts largely used probationary officers and police officers, all districts also incorporated more-senior officers onto the PILOT teams,

---

[2] Again, we assumed that CRU unit man-hours were consistent each month for control districts.

[3] Total hours divided by 7 (months), divided by 22 (days per month), divided by 3 (districts).

Table A.2
Total Hours Worked in Control Versus Treatment Groups, June–December 2012

Hours Worked	Experimental Group				Control Group
	District A	District B	District C	Total	Total
By job class					N/A
Probationary police officer	667	83	198	948	
Officer	1,503	136	509	2,148	
Corporal I	25	3	6	34	
Corporal II	368	21	54	443	
Corporal III	46	0	243	289	
Corporal IV	469	53	72	594	
Sergeant	626	61	89	776	
Total	3,704	358	1,171	5,233	5,999
Nonsergeant-to-sergeant ratio					
Mean	5	5	11		
Median	5	5	12		

although to different extents. District C employed a strategy of more officers (nonsergeants) to supervisors (sergeant) on average (11 to 1) than the other districts (5 to 1). Each district's strategy in this respect remained fairly stable, as the mean and median nonsergeant-to-sergeant ratios are similar.

In total, the labor expense attributable to implementing the PILOT strategy was approximately $167,650 (see Table A.3). For treatment districts, the most utilized job class was officers, followed by probationary police officers. Officers generated the greatest cost though, followed by sergeants.

The number of potential hours of analysts' and administrators' time devoted solely to implementing PILOT, as opposed to reporting requirements of the grant, was calculated (see Table A.4). Based on reporting in the grant budget, the system administrator, the crime analyst, and records personnel were included. Records personnel devoted almost all their time to administrative duties, except filling in FI card data. Given the large uncertainties about the possible proportion of time spent on implementing PILOT, a fairly large range (300 to 900) of analyst and administrative manhours directly attributable to conducting PILOT was estimated. Using actual data on hourly pay, a potential $8,500 to $25,000 spent over the trial period to implement only PILOT activities was calculated.

Table A.3
Labor Cost for Treatment Group, June–December 2012

By job class	Total Hours Worked	Hourly Pay[a]	Total Cost
Probationary police officer	948	$25	$23,532
Officer	2,148	$29	$62,221
Corporal I	34	$32	$1,076
Corporal II	443	$33	$14,712
Corporal III	289	$34	$9,849
Corporal IV	594	$38	$22,495
Sergeant	776	$44	$33,764
Total	5,233		$167,649

NOTE: Any differences are due to rounding.
[a] Overtime rates.

Table A.4
Parameter Assumptions and Values for Supplemental Labor Expenses

Parameter	Value		
Annual hours allotted to PILOT per analyst	600		
Number of working days per year	235		
Average expected daily hours on PILOT for analysts	2.6		
Weighted average overtime hourly rate	$29		
Assumptions about PILOT time spent on implementation (e.g., not reporting)	**Minimum**	**Best**	**Maximum**
Analysts	25%	50%	75%
Records	1 hr/day	2 hrs/day	3 hrs/day
Estimated Hours and Costs			
Total analyst/administrative man-hours	291	582	873
Total costs	$8,461	$16,922	$25,383

Since the information on the number of hours by job class in the control districts was not recorded, a series of assumptions to estimate the labor cost in the control group was made. Initially, we assumed a nonsergeant-to-sergeant ratio of 5:1 (the mean of treatment districts) for control districts, and the relevant corresponding average hourly pay was applied. As seen in Table A.5, this resulted in a potential total labor cost of $195,478 to $204,600, with a best estimate of $202,455.

Vehicle Costs

During the intervention, officers conducted special operations in hot spot (the control group) or predicted crime spike areas (the treatment group). The special operation activities involved patrol, stops and searches, and collection of local information from the public; all of these required the use of police vehicles.

Data from the Federal Highway Administration estimate that the average annual vehicle miles traveled by police for 2011 is 15,160. Using this along with the assumption that a police vehicle is used every day of the year for an eight-hour shift, on average, a police vehicle travels five miles for each hour worked (American Public Transit Association, 2011). Vincentric (2010) estimates that police vehicle expenses per mile are $0.55, including depreciation, financing, fuel, maintenance, insurance, repairs, and

Table A.5
Labor Cost Estimates for Control Group, June–December 2012

	Hours Worked	Potential Hourly Pay[a]	Total Cost
Minimum			
District officers for property operations	3,310.7	$33	$107,885
CRU special emphasis operations	2,688.0	$33	$87,593
Total	5,998.7		$195,478
Best estimate			
District officers for property operations	3,310.7	$34	$111,736
CRU special emphasis operations	2,688.0	$34	$90,720
Total	5,998.7		$202,455
Maximum			
District officers for property operations	3,310.7	$34	$112,919
CRU special emphasis operations	2,688.0	$34	$91,681
Total	5,998.7		$204,600

SOURCE: Based on SPD data. Authors calculated pay and total cost.

NOTE: Any differences are due to rounding.

[a] Weighted average.

Technical Details 59

opportunity cost. The next assumption was that only ranks between probationary officer and corporal were in the field. Given reports of altering numbers of person-vehicle teams of one person, one car and two people, one car, a rate of 1.5 people per car was applied. The relevant number of man-hours in each group was then applied to the number of miles traveled per hour and the cost per mile to calculate total vehicle cost per group.

Results may be sensitive to the number of miles traveled in an hour. Therefore, our initial assumptions were that the treatment and control groups travel the same number of miles per hour worked and that both employ a strategy of using only non-sergeants in the field (best estimate). For a minimum estimate, we assumed that non-sergeant officers travel 20 percent less than the average (one mile less traveled per hour). For a maximum estimate, we assumed that officers travel 20 percent more than the average (one mile more traveled per hour), and that all officers, including sergeants, travel when working on property crime–related special operations.

Results of the vehicle cost estimations, shown in Table A.6, indicate that the control group potentially spent between $7,000 and $13,200, and the treatment group spent in the range of $6,500 to $11,500 on vehicles involved in property crime–related special operations over seven months.

Technical Details of Prediction Model Evolution

To forecast crime risk for the grid cells, the team experimented with three types of models: RTM, logistic regression, and a combination of the two. The size of the grid squares selected to test each model, 400-square-foot grid sizes, was based on guidance

Table A.6
Vehicle Cost Calculations for Treatment and Control Groups, June–December 2012

	Patrol Hours Worked	Average Hourly VMT	Vehicle Cost per Mile	Person per Vehicle	Total Cost
Minimum					
Control group	4,799	4	0.55	1.5	$7,038
Treatment group	4,457	4	0.55	1.5	$6,537
Best estimate					
Control group	4,799	5	0.55	1.5	$8,798
Treatment group	4,457	5	0.55	1.5	$8,171
Maximum					
Control group	5,999	6	0.55	1.5	$13,197
Treatment group	5,233	6	0.55	1.5	$11,513

from the RTM team at Rutgers,[4] which recommended that grid sizes be roughly the size of a single block. Each model used the following risk factors to predict risk in each grid cell: persons on probation or parole, prior six months of tactical crime, prior 14 days of tactical crime, prior six months of disorderly calls for service, prior six months of vandalisms, and presence of at-risk buildings that had seen large numbers of calls for service. Factors tested but not used included the juvenile population, presence of alleys, presence of schools, previous six months of juvenile arrests, and known drug houses.

RTM uses the following process to estimate risk:

- The analyst uses the tool to identify correlations between crime risk and the presence of a number of hypothesized geospatial risk factors.
- For the most-correlated factors, the tool then counts the number of factors present in each cell, and highlights cells with more than a certain number of factors present. The more factors present, the redder the color. In the case of the SPD, cells with two factors present were colored yellow; cells with three to four factors present were colored orange; and cells with five to six factors present, the most possible, were colored red.

For the logistic regression method, the SPD used a forward stepwise approach to select the model of best fit to forecast probabilities of tactical crimes within grid cells in a month. The regression model was rerun monthly to account for the most-recent month's data; it was never retrained.[5] The resulting crime maps were geospatial cells highlighted in orange and red if the cells had probabilities of 40–60 percent and 61+ percent, respectively, of experiencing at least one tactical crime over the next month.

---

[4]   See the website for the Rutgers Center on Public Security: http://rutgerscps.weebly.com/.

[5]   This was based on advice provided to SPD staff by other agencies and vendors.

References

American Public Transit Association. 2011. *2011 Public Transportation Fact Book*. Washington, D.C.

Baranowski, T., and G. Stables. 2000. "Process Evaluations of the 5-a-Day Projects." *Health Education and Behavior*, Vol. 27, No. 2, pp. 157–166.

Bertrand, M., E. Duflo, and S. Mullainathan. 2004. "How Much Should We Trust Differences-in-Differences Estimates?" *The Quarterly Journal of Economics*, Vol. 119, pp. 249–275, DOI: 10.1162/003355304772839588.

Bowers, K., S. Johnson, R. T. Guerette, L. Summers, and S. Poynton. 2011. "Spatial Displacement and Diffusion of Benefits Among Geographically Focused Policing Initiatives." *Campbell Systematic Reviews*, Vol. 3.

Braga, A. A., A. V. Papachristos, and D. M. Hureau. 2012. "The Effects of Hot Spots Policing on Crime: An Updated Systematic Review and Meta-analysis." *Justice Quarterly*, DOI: 10.1080/07418825.2012.673632.

Caplan, J. M., Kennedy, L. W., eds. 2010a. *Risk Terrain Modeling Compendium*. Newark, N.J.: Rutgers Centers on Public Security.

———. 2010b. *Risk Terrain Modeling Manual: Theoretical Framework and Technical Steps of Spatial Risk Assessment for Crime Analysis*. Newark, N.J.: Rutgers Centers on Public Security.

Caudill, J. W., R. Getty, R. Smith, R. Patten, and C. R. Trulson. 2013. "Discouraging Window Breakers: The Lagged Effects of Police Activity on Crime." *Journal of Criminal Justice*, Vol. 41, pp. 18–23.

Eck, J., S. Chainey, J. G. Cameron, M. Leitner, R. E. Wilson. 2005. *Mapping Crime: Understanding Hot Spots*. NCJ 209393. Washington, D.C.: National Institute of Justice.

Levitt, S. D. 2002. "Using Electoral Cycles in Police Hiring to Estimate the Effects of Police on Crime: Reply." *American Economic Review*, Vol. 92, pp. 1244–1250.

Linnan, L., and A. Steckler. 2002. "Process Evaluation for Public Health Interventions and Research: An Overview." In *Process Evaluation for Public Health Interventions and Research*, ed. L. Linnan and A. Steckler, pp. 1–24. San Francisco: Jossey-Bass.

Marvell, T. B., and C. E. Moody. 1996. "Specification Problems, Police Levels, and Crime Rates." *Criminology*, Vol. 34, pp. 609–646.

Office of Justice Programs. No date. "Tips for Using CrimeSolutions.gov." CrimeSolutions.gov, National Institute of Justice. As of March 31, 2014:
http://www.crimesolutions.gov/about_tips.aspx

Perry, W. L., B. McInnis, C. C. Price, S. Smith, and J. S. Hollywood. 2013. *Predictive Policing: The Role of Crime Forecasting in Law Enforcement Operations.* RR-233-NIJ. Santa Monica, Calif.: RAND Corporation. As of May 1, 2014:
http://www.rand.org/pubs/research_reports/RR233.html

Reno, S., and B. Reilly. 2013. "The Effectiveness of Implementing a Predictive Policing Strategy for Tactical Operations." Paper presented at the 2013 International Association of Chiefs of Police Conference, October 21, Philadelphia.

Stroud, M. 2014. "The Minority Report: Chicago's New Police Computer Predicts Crimes, but Is It Racist?" *The Verge*, February 19. As of March 31, 2014:
http://www.theverge.com/2014/2/19/5419854/
the-minority-report-this-computer-predicts-crime-but-is-it-racist

Vincentric. 2010. "A Look at Lifecycle Costs for Law Enforcement Vehicles." As of May 12, 2014:
http://vincentric.com/Portals/0/Market%20Analyses/Law%20Enforcement%20Lifecycle%20
Cost%20Analysis-%20Prepared%20Feb%202010.pdf